W9-BAH-155

SAVING
SCIENCE CLASS

SAVING SCIENCE CLASS

Why we need hands-on science to *engage* kids, *inspire* curiosity, and *improve* education

CHRIS McGOWAN

Prometheus Books
59 John Glenn Drive
Amherst, New York 14228

Published 2017 by Prometheus Books

Inquiries should be addressed to
Prometheus Books
59 John Glenn Drive
Amherst, New York 14228
VOICE: 716–691–0133
FAX: 716–691–0137
WWW.PROMETHEUSBOOKS.COM

21 20 19 18 17 5 4 3 2 1

Library of Congress Cataloging-in-Publication Data

Names: McGowan, Christopher.
Title: Saving science class : why we need hands-on science to engage kids, inspire curiosity, and improve education / by Chris McGowan.
Description: Amherst, New York : Prometheus Books, 2017. | Includes index.
Identifiers: LCCN 2016035799 (print) | LCCN 2016040792 (ebook) | ISBN 9781633882171 (hardcover) | ISBN 9781633882188 (ebook)
Subjects: LCSH: Science—Study and teaching. | Science—Study and teaching—Great Britain. | Motivation in education. | Motivation in education—Great Britain.
Classification: LCC Q181 .M1285 2017 (print) | LCC Q181 (ebook) | DDC 507.1—dc23
LC record available at https://lccn.loc.gov/2016035799

Printed in the United States of America

To Liz with Love—there as always

CONTENTS

ACKNOWLEDGMENTS

A decade ago *Time* magazine, to which I have subscribed for eons, ran a cover article titled "Is America Flunking Science?" The cover depicted a schoolboy, lab coat blackened by an obviously disastrous experiment. This illuminating article is acknowledged as the spark that eventually led to my writing of this book.

First and foremost, I thank Steven L. Mitchell for his enthusiastic response to my proposal to write this book, and for his subsequent acquisition of the manuscript. As my editor, he gave valuable input and direction from the outset, helping me keep the narrative on course. His patient guidance through some of the nuances of the house style was very much appreciated; so too was his timely tuition on footnoting. Thank you, Steven, for this and more.

Of all my previous books, this was by far the most arduous and demanding, primarily because of the subject matter. Owing to the contentious nature of some aspects of the work, I sought the opinions of some of my peers on certain chapters. For their valuable comments and helpful discussions, I thank professors Jaime Alvarado, Judy Massare, Ryosuke Motani, and Jeffrey Thomason. I am deeply grateful to each one of you for your input and much appreciated support.

Several other highly respected academics have taken time from their busy schedules to review advanced reading copies of this book and make commentaries on its content. For their generous support and much appreciated words, I thank William C. Burger, Peter Dodson, Taner Edis, Stephen Gatesy, Quanyu Huang, Colin Pask, Frank Ryan, and Scott Sampson; Judy Massare is also thanked here too.

One unexpected bonus of publishing with Prometheus Books has been working with such a dedicated and accommodating group of professionals. These committed people went out of their way to help me through every stage of the project. One of the first was Hanna Etu, a wellspring of important information, from copyright regulations and sources of illustrations, to giving feedback on certain things I wrote. Hanna was a steadying hand on the tiller, always available when needed—the cheerful voice at the end of the phone; the respondent to my emails, almost as soon as I had pressed the send button. I later learned that she had also emended all my endnotes, so as to conform to the house style. I am truly grateful for all her support.

Cate Roberts-Abel, who first assisted me with important feedback on the quality of some of my illustrations, helpfully steered me through the production process. Her much appreciated advice and assistance included giving invaluable insights into indexing, providing me with informative reading material and exemplars of indices from other books. Cate's timely resolution of some unforeseen problems in the advanced reading copy saved me from a good deal of trouble and angst.

I was exceedingly fortunate to have someone of such meticulous care as Sheila Stewart as my copyeditor—nothing slipped past her sharp eyes. I am sincerely grateful to her for the errors she caught, the facts she checked, the points she raised, and for the numerous improvements that resulted from her close attention to detail. I also thank her for the practical assistance she gave me throughout the editing process, with valuable discussions, opinions, and advice—I'll miss our phone calls. Her genial forbearance of my many minor emendations during the eleventh hour is also greatly appreciated.

Cheryl Quimba, my publicist, has been unstinting in the help she has given me in my attempts to bring this book to people's attention. I thank her for critically reading drafts of proposals I have written, and for giving valuable feedback and advice for pursuing certain avenues of potential interest. I am also grateful to her for following up on some of my suggestions.

Early on in the process I had the pleasure of working with Mark Hall, the communications editor, who drafted the catalog description of the book. Much impressed by his distillation of the essence of my work, I had few contributions to make to his encapsulation of the content.

Freelancer Jeff Schaller designed the front cover, with in-house assistance from Jackie Cooke, who worked on the spine and back cover. The book was designed by Bruce Carle, Prometheus's typesetter. Jeff Curry gave last-minute help, for which I am truly grateful. My thanks to you all.

I thank Jim Almond of Shropshire, England, for his magnificent photographs of an Arctic tern and of an osprey. My thanks also to Duade Paton of New South Wales, Australia, for his superb image of an albatross.

For her generous help in obtaining high-resolution images of the

Wright brothers' aircraft and the facsimile of their wind tunnel I thank Stephanie Kays of Denison University, Granville, Ohio, formerly the archivist at Wright State University, Dayton, Ohio. For their assistance in obtaining an image of Newcomen's engine I thank Mark Bainbridge and Renée Prud'Homme of Worcester College, Oxford, and Jen Burford and Samantha Sherbourne of the Bodleian Libraries, Oxford.

My sincere thanks to Richard Berman and his fellow science teachers at Newmarket High School, Ontario. These dedicated teachers gave so generously of their time and hospitality, and welcomed me into the science lab. I also thank the students for making me feel so completely at home. Witnessing the mutual respect, enthusiasm, and camaraderie within that lab was an inspiring experience.

Steven L. Mitchell will attest to the computer problems I encountered during the writing of this book. Worst among these was a malware infection, occurring at a most critical time. For resolving that issue, and many more beside, I thank the computer wizards at easyTechCare Inc, North York, Ontario, whose services I have been using for many years. I would also like to thank the very helpful staff at my local Staples store in Aurora, Ontario, for everything from image scanning and printing, to ordering certain glues not available on their shelves, and resolving a terminal printer problem.

Working with the incomparably talented Julian Mulock on yet another book was a great joy. Julian has been illustrating my scientific endeavors since our early days of working at the Royal Ontario Museum, a tradition that continued long after we both left. Because of prior commitments, he was unable to commence work until perilously close to the illustration deadline. Unfazed, he completed the task with the professional acumen I have come to admire over all the years. I thank him for his splendid drawings that enliven the pages that follow.

Liz, my wonderfully supportive wife, has been the paragon of patience and understanding throughout my self-imposed year of withdrawal from normal life. While ensconced in my office from pre-dawn to post-dusk, she kept me isolated from everything unrelated to working on this book. She also read everything I wrote, giving valuable comments, criticisms, and constructive advice, which were of immeasurable benefit. No words can express my gratitude for this, and for everything else. We are indeed fortunate to share our life together.

INTRODUCTION

I have been a scientist most of my life, my curiosity having been aroused at an early age. By the time science was being taught as a separate subject at school, I was completely captivated. Part of the reason for my interest came from having had good teachers, who shared their enthusiasm and knowledge of science with their students. Inspired by what I experienced in the classroom, I conducted experiments at home, learning through doing. Growing up during the shortages of post-war England encouraged creativity, and my experiments involved improvising with things found around the house. Much of the laboratory equipment at school was quite basic too, for similar reasons—expensive apparatus is not a prerequisite for doing science.

Many of my classmates were as interested in science as me, and I was not the only one to pursue a career as a scientist. Today, it seems, this is no longer the case, as reported in a recent editorial in the *New York Times*:

> Nearly 90 percent of high school graduates say they're not interested in a career or a college major involving science. . . . One of the biggest reasons for the lack of interest is that students have been turned off . . .[1]

Before taking early retirement, I had neither reason nor time to find out how science is being taught in schools, nor to investigate why youngsters are losing interest in the subject. However, during the last few years, I have spent time in school classrooms, both as an observer and as a teacher. I have also been talking with teachers and listening to their concerns about teaching science according to current directives. Much time has been devoted to wading through pages upon pages of educational publications and documents—including *A Framework for K–12 Science Education* and the Next Generation Science Standards.[2]

The reason why students are turned off science at school is abundantly clear; so too is what is required to reverse this. My findings will be roundly rejected as highly contentious in certain educational circles, so I should say something about my background and why I should be taken seriously.

On leaving school, I studied for a BSc honors degree in zoology. After graduating, I became a full-time schoolteacher, instructing students aged between twelve and eighteen at three fundamentally different types of schools. My first appointment was to a technical school, similar to the one I had attended from the age of thirteen. The second post was to a secondary modern school, the same category of school that I had attended for two years, before being elevated to the technical school. My last post was to a grammar school, the highest level of learning, where students could only gain admission by passing their eleven-plus exam,[3] something I had failed to accomplish when I was eleven.

I loved teaching but did not want to stay at school forever, so I enrolled as a part-time graduate student at London University, studying paleontology. Receiving my PhD three years later, I was remarkably fortunate to obtain a curatorial position in vertebrate paleontology at the Royal Ontario Museum in Toronto, one of Canada's premier museums. I was later cross-appointed as an assistant professor of zoology at the University of Toronto.

During my academic career, I published forty-six papers in major scientific journals, along with two text books and six popular books on science. My research was primarily on ichthyosaurs, a group of marine reptiles contemporaneous with dinosaurs. Among my contributions to the field was the naming of three new genera and eight new species.[4] I received continuous federal funding for my research for twenty-seven years, during which time I was recognized for "25 years of excellence for important research achievements" by the Natural Sciences and Engineering Research Council of Canada.[5] I was also awarded Honorary Membership by my professional body, the Society of Vertebrate Paleontology. I attained the rank of full professor at the University of Toronto and Senior Curator at the Royal Ontario Museum.

Teaching has always been an important part of my academic life, and one of the challenges that captured my interest was explaining complex ideas in intelligible terms, to make science accessible. Science is a practical subject and forty years of teaching at all levels—from high-school students to graduate students—has convinced me that the best way for students to learn is through their own experiences. Devising experiments for students to conduct themselves has therefore been an integral part of my teaching practice. Having written the book *Make Your Own Dinosaur out of Chicken Bones*, I think there are few areas of school science where

I could not create a suitable experiment for youngsters to learn through hands-on experience.

Schools have limited budgets for purchasing equipment and supplies, which may restrict the amount of experimentation in the classroom. To illustrate what can be done with limited resources, I staged a hands-on science week at the Royal Ontario Museum, during school spring break in 2010. Using everyday items, I explored various topics—from demonstrating air density by using atmospheric pressure to collapse empty milk cartons, to "rubberizing" leftover chicken bones to show the composite nature of bone.

Interacting with enthusiastic youngsters and their parents after each presentation revealed how few students conducted experiments at school.

When I started writing this book I was well aware of the serious shortcomings in the teaching of science in American and Canadian schools. However, I was unprepared for some of the things I discovered, and suspect my readers will be too. Given the serious environmental problems facing our world, scientific literacy has never been more important. Yet we live at a time when more and more people know less and less about science. This situation can only be rectified in the school classroom, but this requires returning responsibility for what is taught in science to those who are scientifically literate themselves.

I have been exceedingly fortunate in the things I have seen and done during my scientific career, and in my diverse teaching and learning experiences. For me, science and teaching are engaging pursuits, and I seek to entertain as well as to inform my readers with a narrative liberally seasoned with personal anecdotes.

My aspirations for this book are that it will provoke change, empowering teachers to reinstate *real* science in the classroom.

CHAPTER 1

FROM CRUSHED CANS TO CURRICULUM CONSULTANTS

Mr. Jordan captured my attention back in the early fifties, along with the attention of all the other twelve-year-old boys in my class. Taking an empty gallon can, he poured in a splash of water and then placed it on a tripod over a Bunsen burner. Once steam was billowing, he removed the can from the flame, quickly screwed on the lid, and set it down on the bench.

Nothing happened.

We waited patiently, but still nothing happened. The seconds ticked by and there were some whispers from the back of the class. And then it happened. With a loud metallic crunch that made everyone jump, the can was crushed almost flat. Holding up the can for all to see only added to our wonder—it looked as if it had been run over by a double-decker bus.

More than half a century later and I was captivating youngsters with science myself. The venue was the Royal Ontario Museum in Toronto, where I was staging a hands-on science extravaganza during the spring-break holiday. It was here, at Canada's premier museum, that I had enjoyed a magical career as dinosaur curator and professor of zoology. Having taken early retirement to write more books and to free up a position for younger blood, I was spending a week engaging children in science.

I began with the tin-can experiment, receiving all the shrieks and gasps, just as Mr. Jordan had. "Can anyone explain what made this happen?" I asked, and was not surprised when nobody could. I described how the steam had displaced all the air from the can, so screwing on the lid had captured a tin full of hot steam. As this cooled, it condensed into tiny water droplets. These occupied such a small amount of space that the can was now essentially empty. "What do we call such an empty space?" Nobody volunteered "a vacuum." After explaining how the can was flattened by air pressure—15 pounds per square inch—I asked what caused the pressure. With no correct responses, I explained it was the weight of all the air pressing down from above. "Just think of all those thousands of feet of air, reaching way up into the sky." I glanced up at the ceiling. "It's like the water pressure you feel in your ears when you dive down in the deep end of a pool.

"How much does a cubic meter of air weigh?" Holding up a meter rule, I measured out an imaginary cube in the air.

"Does it weigh this much?" I asked, holding up a feather. Lots of heads started shaking.

"Not as much!" someone shouted.

"Actually, it weighs more," I replied and exchanged the feather for a mini-packet of rice cereal. "How about this much?" The packet weighed 21 grams (3/4 ounce).

Some heads were shaking, but others were nodding. They were getting the idea.

"It's *lots* heavier," I said and held up a regular-sized packet of the cereal. "Does a cubic meter of air weigh this much?" Checking the label, I read out its weight: 11 ounces, or 305 grams.

A few heads were nodding.

"How about this?" I held up a tin of coffee weighing 370 grams. "Is a cubic meter of air this heavy—just over three-quarters of a pound?"

Nobody thought so.

"Actually, the air weighs a little more than this one-liter carton of milk." I held the full carton above my head. "This milk weighs a kilogram and a cubic meter of air weighs 1.2 kilograms—that's just over two-and-a-half pounds!"

Many people were finding it hard to believe that the density of air was 1.2 kilograms per cubic meter, so I told them I could demonstrate this for them after the show, with a simple experiment using a helium balloon.

"Okay. How many youngsters would like to try the can-crushing experiment?" There was a sea of waving arms.

"Tin cans and Bunsen burners are hard to come by these days, but you can use an empty milk carton and a microwave instead."

After adding a miniscule amount of water to an empty milk carton, I gave it to one of the youngsters crowded on the stage. Following my instructions, she placed it inside the microwave. Then, with the dial set for a minute at full power, she pressed the button and the countdown began.

When time was up, I opened the door, quickly screwed on the plastic cap—good and tight—and stood the carton on the table.

The speed at which the carton collapsed compensated for the absence of a tin-can crunch, and the youngsters watched in wonder as the sides sucked in and the carton leaned further and further to one side. Seconds later it toppled over, as if dead.

"We could have made that happen even faster," I told them. "Do you want to see how?" Of course they did.

This time, after removing the hot carton from the microwave and screwing on the cap, I stood it on a plate. "Watch this," I said, dousing the carton with a cup of cold water. The carton immediately collapsed. "That's because the cold water instantly condenses the steam into droplets."

They all wanted to try this for themselves, but I had to keep an eye on the time.

"Crushing cans and cartons is lots of fun, but there's a practical side to this experiment." I went on to explain how air pressure was used to power the world's very first engine, over three hundred years ago. Images then flashed onto the screen of a working replica of Thomas Newcomen's atmospheric engine. One of the photos showed me, standing above the eight-foot-high cylinder, to give an impression of its massive size. Leaning over, I am looking down at the top of the piston—it is broader across than my shoulders.

This engine worked by filling the cylinder with steam and then squirting cold water inside to condense it. The partial vacuum that formed caused the air pressure to push the piston down to the bottom. The piston was connected, by a chain, to one end of a huge overhead beam. The other end of the beam was attached to a heavy counterweight. Pivoted at its center, the beam could rock up and down, like a giant teeter-totter. As the air pressure pushed the piston down the cylinder, depressing one end of the beam, the other end was raised, lifting the counterweight. Once the piston reached the bottom of the cylinder at the end of the power stroke, the counterweight caused the beam to rock back the other way, pulling the piston back toward the top. As this return stroke began, a valve opened at the bottom, releasing steam into the cylinder. Once the piston reached the top and the cylinder was full of steam, the valve closed and a second one opened, spraying in cold water for another power stroke.

Before Newcomen's revolutionary invention, the only power available in the world came from wind, water, and from human or animal muscle.

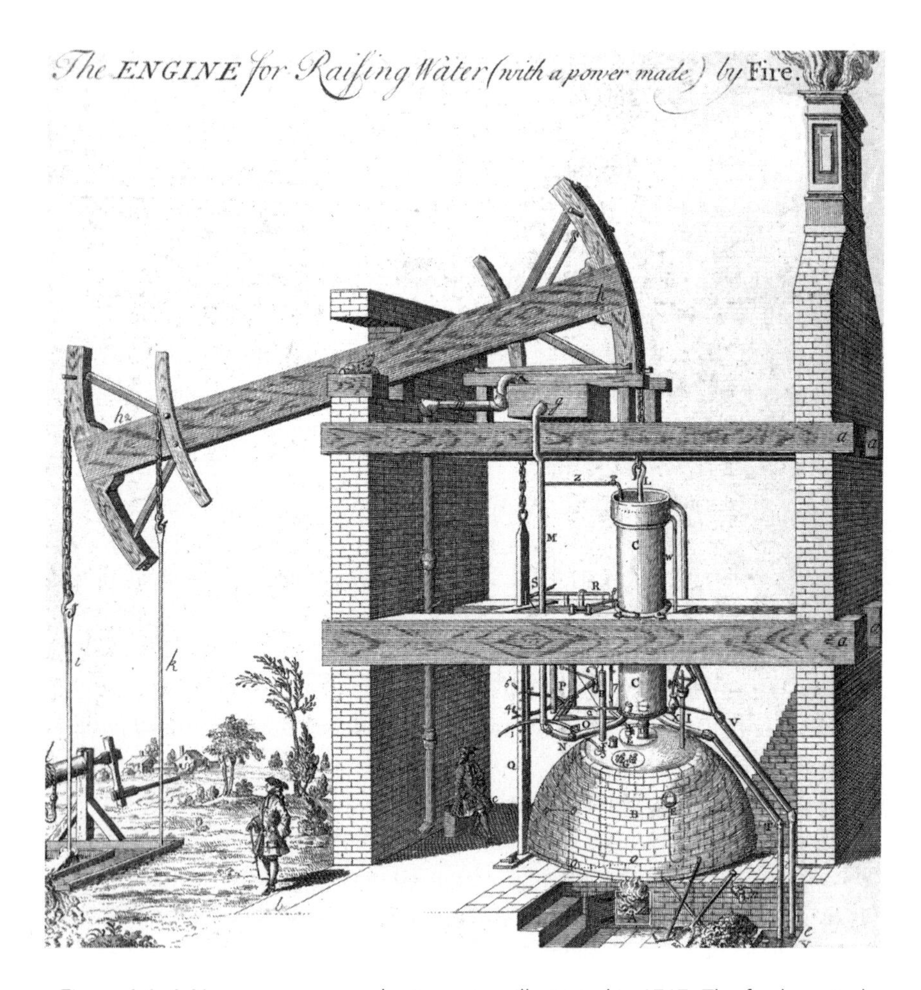

Figure 1.1: A Newcomen atmospheric engine, illustrated in 1717. The frock-coated gentleman on the left of the wall gives an idea of its immense size.

From air pressure and density, I turned to the fluid properties of air. Most people think of a fluid as something that comes out of a bottle, but gases are fluids too. And an important property that gases and liquids share is the ability to flow. Explaining about airflow was my introduction to how things fly.

I explained that birds, bats, and Boeings have one thing in common: wings with a convexly curved upper surface. Using both hands, I picked up a sheet of paper by its bottom corners, pointing the lower edge down toward the floor so the free edge arched away from me in a gentle curve.

Then, holding the nearest edge up to my lips, I blew air as hard as I could across the curved upper surface. The sheet instantly popped up, due to the reduction in air pressure over the convex surface. Structures like this, which generate lift by virtue of their convexly curved upper surfaces, are referred to as airfoils.[1]

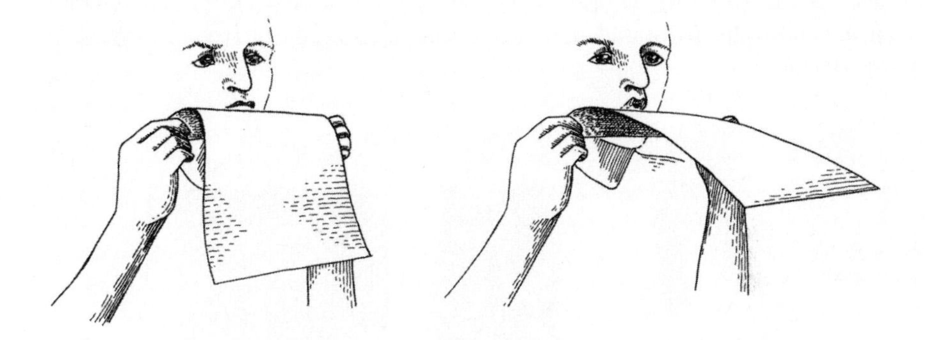

Figure 1.2: Blowing air hard across the curved upper surface of a sheet of paper causes it to pop up.

The next airfoil was more up-market: the inside tube from a toilet paper roll. I had cut the tube in half lengthwise and tethered the four corners to a rectangle of card, using equal lengths of thread. There was enough slack in the threads to allow the airfoil to rise by a couple of inches. Using a hair-dryer, I made the airfoil takeoff and fly. All the youngsters wanted to try, and some of their parents too.

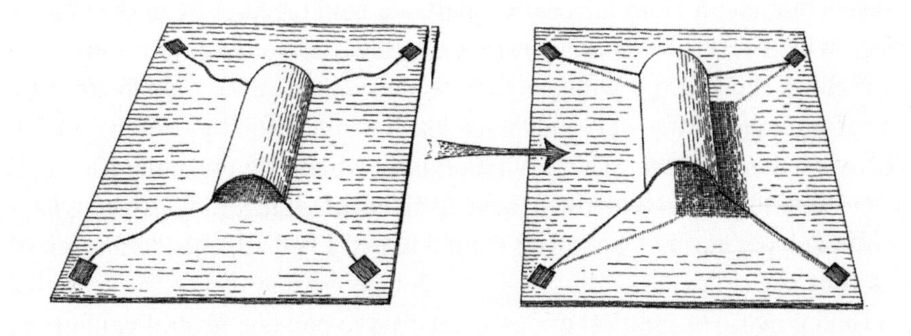

Figure 1.3: A simple airfoil, made from a cardboard tube cut in half, before and after being blown by a hairdryer.

I had an assortment of natural airfoils to show my audience. These ranged from the wing of a dead bird that I had found on my street to some bird wings borrowed from the museum's ornithology collection. Several youngsters joined me on stage for a closer look, and I handed the oldest one an eagle wing to hold. The wing was longer than her own arm, and she was as impressed by its lightness as she was by its size. One of the reasons it was so light, I pointed out, was because so much of it was formed of feathers.

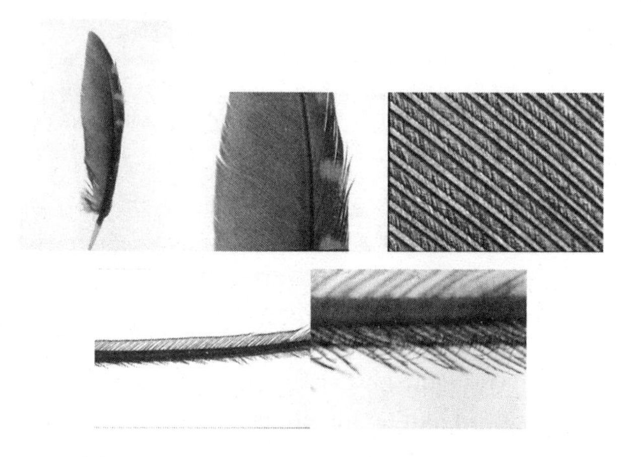

Figure 1.4: A feather, viewed at increasing magnifications. Notice the minute hooklets in the last photo.

By projecting images of a feather onto the screen at increasing magnifications, I explained its remarkable lightweight construction. The barbs, which branch off from the central shaft, are held together by minute hooklets, which act like Velcro. This gives stiffness and strength for minimum weight. I showed the youngsters how they could ruffle up a feather by stroking it the wrong way, and then make it whole again by stroking it back between their fingers. "That's what birds do when they preen themselves."

Wing feathers, I explained, have the form of an airfoil, with a convexly curved upper surface. I showed them one that I had attached to a piece of card by the tip of its quill, using scotch tape, so that it was free to hinge up and down. The gentlest of puffs caused it to pop up. Several youngsters tried this for themselves.

Returning to the eagle wing, I held it up against the shoulder of a small lad. "So, if you had a large enough pair of wings, do you think you could fly?" Feeling lucky, he said he could, but the others disagreed. When I asked them why they would be unable to fly, even with the biggest of wings, one volunteered that it was because we are too heavy. "Good answer," I commended, and began comparing the lightweight construction of bird bones with those of non-flying animals, like ourselves. I showed them various arm bones that had been sawn through longitudinally, so they could see the relative thickness of their walls. All were hollow, and those of the birds were so very much thinner and lighter. "You can make your own bone specimens using leftovers from dinner," I suggested. Holding up a small hacksaw, purchased from my local hardware store, I showed them some chicken and lamb bones that I had sawn in half lengthwise.

"Aircraft are built using similar lightweight construction."

As the image of a Boeing 787 appeared on the screen, I explained how additional weight reduction was achieved in passenger jets by using composite materials of plastic and carbon fiber. These composites are much lighter than metals, and Boeing achieved a twenty percent weight reduction by using them in the 787.

"Can anyone think of any composites that we use?" Nobody could, so I asked if anyone had ever broken an arm or a leg and had it set in a plaster cast. There were a few nods among the adults. "Plaster casts are composites of plaster and bandage. We use similar casts for wrapping up fossils in the field," I said, holding up a field jacket containing a dinosaur bone. "But we use burlap instead of bandage. The advantage of composites like this is that they are stiff and strong without being too heavy."

I asked the youngsters whether they had ever made papier-mâché at school. This simple composite material is made by dipping strips of newspaper into a flour and water paste and building up layers that are then left to dry. Seeing some blank faces, I showed them a quick and easy way of doing this, using strips of paper towel and white glue. Some of the youngsters tried tearing a sample I had made from only six layers; they were surprised at how tough and stiff it was.

Composite materials date back at least to ancient Egyptian times, and an image of a three-thousand-year-old mud-brick temple flashed onto the screen. The bricks, made by mixing mud with straw, are still used today, and I showed my audience a specimen that I had collected from the Egyptian desert.

"Bone is a composite material too," I said, picking up a long bone.

"It's a composite of a calcium mineral, which makes it stiff and hard," I tapped the bone on the table, "and a fibrous protein, called collagen, which gives a degree of springiness." Grasping the bone by both ends, I made a flexing motion. "Together they form a tough, lightweight material, far stronger than either component is on its own." Laying down the bone, I picked up a much smaller one—the femur (thigh bone) of a chicken.

For several weeks, I had been saving leftover chicken bones, boiling and scrubbing them clean for this week's event. Holding the bone above my head, I threw it down hard onto the floor, to show how tough it was. Next, I tried breaking it with my hands. The shaft of the finger-length bone was half the width of my small finger but, no matter how hard I tried, I could not break it. Casting around for someone big and strong, I asked if anyone would like to try their luck. The athletic looking fellow who took up the challenge was determined to snap the puny bone, and I began to wonder whether he might succeed. However, after a minute or so of determined effort, he had to admit defeat.

"There's a simple experiment you can do at home to show that bone is a composite material," I began. "It also shows the nature of its two components."

I picked up another chicken femur. It looked the same as the first, except it was darker in color. That was because I had popped it into the oven for an hour or so, alongside the Sunday roast. I tapped it gently against the table, showing it was stiff and hard. Searching around for a suitable volunteer, I handed it to one of the smallest members of the audience and invited her to try and break it. Looking uncertain, she took the bone in both hands and, to her complete surprise, snapped it like a cracker.

Returning to the stage, I picked up another baked bone, explaining how the heat of the oven had broken down the collagen, leaving behind the hard but brittle bone mineral.

It was so brittle that I could break off pieces with my fingernail. When I dropped it onto the floor, it shattered into small pieces.

Fishing inside a small plastic bag, I removed another chicken femur and dabbed it dry. This one had been kept in vinegar for a couple of weeks, which, I explained, had dissolved away the bone mineral, leaving the resilient collagen. It looked exactly the same as an untreated bone. But then, to everyone's surprise, I bent it back and forth as if it were made of rubber.

Collagen, I explained, is found elsewhere in the body, including ligaments, the tough cords that join bones together, and tendons, the cords that attach muscles to bones. When I suggested that most of the youngsters had probably enjoyed eating collagen, they looked unconvinced. Then I said that when bones are boiled up for a long time, their collagen is converted into gelatin. "Strain off the bones, add some color, flavoring, and sugar, and you've got jelly." I think I may have turned some of them off jelly for life.

Returning to my collection of bones, I picked up a long one that had been cut in half lengthwise and said how the limb bones of most animals were hollow. This accords with an engineering principle for increasing strength. To demonstrate how this works, I introduced them to my bone-breaking machine. This is a simple device that holds the specimen being tested in a horizontal position, by its two ends. Weights are then added to a pan, suspended from its midpoint by a steel loop. The weights are added, one at a time, until the specimen breaks. In engineering terms, the horizontal specimen is being loaded as a beam; if it were vertical, it would be loaded as a column.

The specimens to be tested were a hollow tube and a solid rod, of equal length, made from the same amounts of plaster. The two therefore had the same cross-sectional areas, the difference between them being that the tube had a greater outside diameter than the rod, which imparts greater strength. I began the demonstration with the rod, loading it with a 1 kg (2.2 lbs.) weight. Taking care to lower the additional 1 kg weights gently onto the pan, I continued increasing the load. The rod broke with a satisfying crack and the clatter of weights, at a load of 7 kg. Replacing the rod with the tube, I added the weights, one at a time, until there were five on the pan. "Well, that's it then," I announced. "The tube's obviously stronger than the rod." Then, with a smile, I added, "There's no need to add any more weights and break it, is there?" That, of course, is exactly what the youngsters wanted and, to their enthusiastic urging, I continued adding weights. The tube failed at a load of 12 kg, with a loud crash that made many of them jump—you can watch it on YouTube, along with some of the other experiments.[2]

When I asked how many of them would like to repeat such an experiment at home, there was a chorus of agreement. The experiment I demonstrated to them involved nothing more elaborate than two cans of pop, two drinking straws, the cap from a juice jug, a handful of quarters, and a twist-tie, along with scotch tape, paper, and scissors.

The plastic cap and paper were used to make a cylindrical container for the quarters, which were used as weights, with the twist-tie forming its handle. The straws were the specimens to be tested, and these were supported, at either end, by a can. One of the straws was the regular supermarket kind; the other was one of those extra wide ones used for milkshakes. Taking each straw in turn, it was laid across the two cans, so the gap it straddled was the same each time. After hanging the coin carrier in the middle of the straw, the ends were taped down to the top of the cans to stop them moving. Predictably, the wider straw supported a much larger load.

I ran three sessions at the museum each day, and these invariably ended with a gathering of enthusiastic youngsters wanting to see and hear more. There was much interest from the parents, too, some of whom expressed regret that their children did no practical science at school. Anxious to know the extent of hands-on science in the classroom, I decided to look into the situation.

Having been a science teacher myself, I had some misgivings about approaching a high school in my area and requesting to sit in on a class. Aside from the intrusion of a stranger in the classroom, there was the burden upon the teacher's time. I recall from my own school-teaching days the feeling of having to run as fast as I could just to stay ahead of my classes—there was no time for diversions.

The enthusiastic and dedicated science teachers who welcomed me into their school were unstinting in their help and consideration. The day began, before classes started, in a room adjoining the teaching lab. This served as a science staff room, an office, and a preparation room for the practicals. As it happens, chemistry was the subject that day, an area of science in which I have had no involvement since I was a student myself. Among the experiments was a series of chemical reactions to be carried out, on a small scale, inside test tubes. With final preparations completed, and equipment and supplies laid out, the lab was ready for the students.

Over forty years had passed since I was last in a school science lab, and I was not sure what to expect. The first thing that impressed me was the mature and cooperative demeanor of the students, and the mutual respect shared between themselves and their teachers. Having been introduced to the class as someone who wanted to see how school science was now being taught, the students accepted me as if I were part of the fixtures. My presence had no effect upon them, nor did they mind answering any questions that I was invited to ask them. The students got on with the job in an orderly manner, working their way through the schedule of experiments that they had to complete. The good rapport enjoyed between the staff and students bore testimony to a happy and well-run school.

I left the building at the end of the day feeling much assured that hands-on science was alive and well, at least at this particular high school. Some things were reminiscent of my school teaching days, but there were differences. One thing that struck me—aside from the mandatory health-and-safety goggles—was the way the students handled their equipment. Instead of holding the test tubes at eye-level when they were doing things with them, they kept them down near the bench. I recall the derisive schoolboy comments we made about the way our chemistry teacher held test tubes and flasks up to his face, looking like the proverbial mad scientist. And then he explained how bringing your eyes level with what you were doing was the only way to do things precisely, and it all made sense. Following his example, I passed this on to my own students, and to those I met that day.

A few weeks after my hands-on science extravaganza at the museum, I received a letter from the Science Teachers' Association of Ontario, the largest organization of science teachers in Canada. One of their members had seen my presentation, and I was invited to be a featured speaker at their forthcoming annual conference, in November. Accepting their invitation, I gave the title of my talk: "Hands-on Science: A Practical Guide." This would be a one-hour presentation, and I would have to devise a number of new experiments, so I decided to set aside an entire month for preparation. Fortunately, I did not have to think about it for another seven months, and I returned to the manuscript I was working on.

It should have been so easy. All I had to do was browse through the Ontario curriculum for a segment of science where I could string together a series of related experiments. This proved surprisingly difficult, and little in the curriculum resembled science as I know it. Instead, I discovered a smorgasbord of sociology and misunderstood science, with meaningless flowcharts linking "Fundamental Concepts," "Big Ideas," and the three "Goals" of the science curriculum. Remarkably, the first goal was "to relate science to society." Relating science to society has as much relevance to the teaching of science as the chemical properties of gold has to the teaching of economics. With its organizational charts and tables, goals and expectations, and quotes from committees, curriculum consultants, and sundry other *experts*, the curriculum bore all the hallmarks of decision-making through consensus, with inputs from everyone.[3]

Before delving any deeper, I must point out that Ontario, where almost forty percent of the population of Canada lives, is not unique in having a curriculum so bereft of real science, or sense. Similar problems can be found throughout the United States, as I will show later (see chapters 5 and 7). A common trait they share is the inclusion of principles and concepts from curriculum consultants, science coordinators, and various other educational theorists, some of whom are cited in the Ontario curriculum. The Ontario document therefore serves as a good exemplar of the science being taught in schools across North America.

The second point I need to make concerns the trend in recent years to incorporate technology into the school science curriculum. In the technological age in which we live, it would be difficult to teach science without discussing some of its practical applications. However, the inclusion of technology as a distinct component of the science class reduces the science content of the curriculum. I recall from teaching school science that there was precious little time to cover the relevant material in the allotted time and see no merit in reducing this still further by including technology. More will be said on this later.

Wading through a science curriculum constitutes cruel and unusual punishment that I would not inflict on anyone, far less my readers. However, it is necessary to spend some time delving into the murky depths to see how serious the situation has become in the science classroom.

The introduction to the Ontario curriculum states that there are "three major goals of the science and technology program," namely:

1. to relate science and technology to society and the environment;
2. to develop the skills, strategies, and habits of mind required for scientific and technological problem solving; and
3. to understand the basic concepts of science and technology.[4]

Relating science to society and the environment *and* making this the primary goal of the science program makes absolutely no sense to me. In attempting to justify this action, reference is made to how much science has expanded our understanding of the world and changed our lives. By the same illogical argument, science has revolutionized computers and electronics, so they, too, should be part of the science curriculum.

Later on it is noted that: "The increased emphasis on science, technology, society, and the environment (STSE) within this curriculum document provides numerous opportunities for teachers to integrate environmental education effectively into the curriculum . . . to take students out of the classroom and into the world beyond the school."[5] Environmental studies and taking students on field trips are relevant to biology and merit inclusion, but this does not justify incorporating STSE education into the curriculum. Unfortunately, however, STSE has become a hallowed part of science curricula across North America and beyond.

The origin and development of STSE education can be traced back to the 1970s and is attributable to the efforts of a number of educational specialists, primarily working in departments of education at universities in the United Kingdom, Australia, and North America. Back then, the concept was referred to as STS education, the *E* for *environment* having been added in more environmentally conscious times.[6]

What is STSE education? There appear to be almost as many definitions as there are authorities on the subject. These range from "teaching science content in the authentic context of its technological and social milieu" to "dealing with students in their own environments and with their own frames of reference."[7] Only an educational theorist could advocate incorporating such esoteric concepts into the science classroom—certainly not schoolteachers faced with the task of interpreting and then teaching such abstractions to their students.

A discourse on science, technology, society, and the environment certainly has a place in our world, especially in our current critical times of change. However, STSE should not be part of the school science curriculum, where it would be redundant anyway if science were taught properly. I cannot, for example, visualize discussing air pressure without mentioning the invention of the world's first engine and how this technology revolutionized industry.

The second goal of the curriculum—"to develop the skills, strategies, and habits of mind required for scientific and technological problem solving"—is similarly redundant. The attributes listed are among those that students would be expected to acquire in a properly taught science program. To give an analogy, if I were teaching someone to ride a bike, that would be the goal, not to develop the skills and strategies of maintaining balance and the habits of mind required to solve the problem of avoiding parked cars.

The only relevant goal of the curriculum is the third one, "to understand the basic concepts of science," and this should have primacy. The education I received at school, over half a century ago, gave me a solid foundation in the basic concepts of science. This gave me the ability to solve problems that I would encounter later in life. I can illustrate this with a recent example.

Most people have difficulty believing that a cubic meter of air weighs just over two-and-a-half pounds (1.2 kg). In pondering how to demonstrate this to my museum audience, I recalled from my school physics that the upthrust on a floating body is equal to the weight of fluid that it displaced. I could use this principle—attributed to Archimedes—to estimate the density of air by using a helium balloon.

To measure the upthrust on a helium balloon, I would need to add just enough weight to it to make the thing float, without rising or sinking. To do this, I attached a length of about twenty feet of string to the balloon and let it go, so that the string it carried aloft was heavy enough to halt its ascent before it touched the high ceiling. Carefully cutting off the excess string at floor level left the correct weight attached to the balloon to make it float at one level. The balloon was now neutrally buoyant—neither rising nor sinking—and the upthrust upon it was equal to the combined weight of the string and the balloon's rubber (the negligible weight of the helium can be ignored). This upthrust was equal to the weight of air displaced by the balloon's volume. The volume of the balloon can be approximated by treating it as a sphere and taking its diameter to be the average of its length and breadth. Before weighing the balloon with its attached string, it has to be deflated. This is done by pinching off the end through which it was inflated, nicking the free end with scissors, and gently releasing the gas.[8] The weight of the deflated balloon and string is equal to the weight of air displaced. From the weight of this volume of air, it is a matter of simple arithmetic to calculate the weight of a cubic meter of air.

Following on from the "three major goals," the curriculum introduces the notion of the "fundamental concepts." These are defined as "key ideas that provide a framework for the acquisition of all scientific and technological knowledge. . . . The fundamental concepts . . . are *matter, energy, systems and interactions, structure and function, sustainability and stewardship*, and *change and continuity*." The curriculum tells us that: "As students progress through the curriculum from Grades 1 to 12, they extend and deepen their understanding of these fundamental concepts and learn to apply their understanding with increasing sophistication."[9]

Aside from my difficulty in thinking of matter, or any of the other listed items, as *concepts*, is the question of why they in particular were selected as the framework for acquiring knowledge. It takes a vivid imagination to visualize youngsters deepening their understanding of *change*, for example, and doing this with "increasing sophistication" into the bargain. As if to strengthen the case for their choice of Fundamental Concepts, we learn that: "The fundamental concepts addressed in the curricula for science . . . are similar to concepts found in science curricula around the world."[10] Ignorance, it seems, knows no boundaries.

For readers shaking their heads and wondering where all this concept nonsense comes from, the answer appears to be given in the quotation that introduces the subject of Fundamental Concepts, in the grades 9 and 10 segment of the curriculum:

> ***Change the focus of the curriculum and instruction from teaching topics to "using" topics to teach and assess deeper, conceptual understanding.***
>
> **—Lynn Erickson**[11]

Dr. Lynn Erickson, a leading light in concept-based learning, is a private consultant who describes herself as "assisting schools and districts around the world with curriculum design and instruction."[12] Graduating with a bachelor's degree in education, she taught elementary school in California before earning a master's degree and a doctorate in Curriculum and Instruction and School Administration. She has written several books on concept-based learning, which she defines for her readers:

> Concept-based curriculum and instruction is a three-dimensional design model that frames factual content and skills with disciplinary concepts, generalizations and principles. Concept-based curriculum is contrasted with the traditional two-dimensional model of topic-based curriculum which focuses on factual content and skills with *assumed* rather than deliberate attention to the development of conceptual understanding and the transfer of knowledge.[13]

In the glossary of one of her books she defines *concept* as: "A mental construct that frames a set of examples sharing common attributes. One- or two-word concepts are timeless, universal, abstract, and broad. . . . Concepts may be very broad macro concepts, such as 'change' . . . or they may be more topic specific, such as 'organism.'"[14] Reading incomprehensible passages like these brings Hans Christian Anderson's story of the emperor's new clothes to mind. In this story, the ruler sought to impress his subjects by appearing before them in an elegant new suit. The emperor had been fooled into believing that the suit, purchased from tricksters, only appeared invisible to those who were stupid. And so it was that the emperor appeared stark naked before his loyal subjects.

The last "innovation" to be addressed in the introduction to the Ontario curriculum is that of "big ideas," which begins with the following quote:

> ***Big ideas go beyond discrete facts or skills to focus on larger concepts, principles, or processes.***
>
> **—Grant Wiggins and Jay McTighe**[15]

Dr. Grant Wiggins, who earned his Doctor of Education from Harvard University, was a secondary school English teacher for many years before forming his own educational consulting company. In the latest edition of his book with Jay McTighe—also a former classroom teacher—we are told that

> At the heart of teaching for understanding is the need to focus on unifying ideas and inquiries, not just discrete and disconnected content knowledge and skill. Big ideas reside at the core of expertise. . . . Because big ideas are the basis of unified and effective understanding, they provide a way to set curriculum and instructional priorities.[16]

Wiggins and McTighe's book is inordinately complex with all its templates, modules and charts. They admit that, "If this is your first time planning with UbD (Understanding by Design), you may have found the process demanding . . ."[17] An understatement indeed.

According to the Ontario curriculum,

> "Big ideas" are the broad, important understandings that students should retain long after they have forgotten many of the details of something that they have studied.[18]

As examples of big ideas, attention is turned to Understanding Life Systems (from grade 3). Two of the big ideas here are that "plants are the primary source of food for humans" and "humans need to protect plants and their habitats."[19] To me, big ideas are things like *evolution by natural selection*, *the atomic theory of matter*, and the *law of conservation of energy*, not this sociological trivia. But I'm a scientist, not a curriculum specialist.

A flowchart is given to complete the picture, showing how the fundamental concepts are linked, through big ideas, to the three goals and overall expectations of the curriculum. And so ended the introduction to the curriculum.

Underwhelmed and unenlightened, I skipped ahead, hoping to find a segment of science suitable for stringing together a series of experiments for my presentation to the science teachers. Eventually, in grade 6, I came to a section on flight that looked promising.

The overview to flight begins with a statement on the societal and environmental effects of the technology. The point is then made that students must first learn about the "certain properties of air"[20] that make flight possible: "For example, air takes up space, has mass, expands, and can exert a force when compressed."[21] That these "certain properties of air" pertain to all other matter that I can think of comes as no surprise in such a flawed document.

After a warning to students about the hazards of exploring flying things, a table is given listing two seemingly relevant Fundamental Concepts: *Structure and Function* and *Matter*. The corresponding big ideas here are: "Flight occurs when the characteristics of structures take advantage of certain properties of air." And the equally inane: "Air has many properties that can be used for flight and for other purposes."[22]

Delving deeper, I discovered that some of the issues under Goal 1 (Relating Science and Technology to Society and the Environment) are that: "Crop dusting from planes allows the chemicals to spread quickly . . ." and "The speed and ease of air travel allow quick transportation of organs for lifesaving transplants . . ."[23] The inclusion of such irrelevant trivia in a science curriculum underscores the absurdity of STSE education.

Issues relevant to Goal 2 (Developing Investigation and Communication Skills) include flying "kites and airplanes a safe distance from overhead hydro wires" and building and testing "a flying device (e.g., a kite, paper airplane, a hot air balloon)."[24] The first item, about avoiding hydro wires, is as irrelevant to developing investigation and communication skills as being careful when riding a bike. And, while flying a kite has only a tenuous connection to the mechanics of flight, hot air balloons are completely irrelevant, the mechanism here being *buoyancy*, not flight.

As part of the same goal, students are expected to "use appropriate science and technology vocabulary . . . in oral and written communication."[25] The glossary provided at the end of the document includes some of these terms. On checking the relevant entries, I found the following definition for *airfoil*: "A teardrop-shaped or nearly teardrop-shaped structure that produces a force or lift as it moves through air; aircraft wings and propeller blades are examples of airfoils."[26] The teardrop-shaped structure to which they refer is a *streamline* not an airfoil (see chapter 4). The streamline shape is used for reducing drag not for generating lift.[27] While most modern aircraft do have wings with cross-sectional shapes approaching a streamline—a strategy for reducing drag—this is not so for supersonic ones.[28]

I found many more incorrect and imprecise definitions in the glossary and began compiling a list. Here is a small sampling:

Solution: "A homogeneous mixture of two or more substances, which may be solids, liquids, gases, or a combination of these."[29] According to this definition, air, a mixture of gases, would erroneously be a described as a solution.

Light: "Radiative energy that can be detected by the human eye and makes things visible . . ."[30] This anthropocentric definition excludes ultraviolet light, which cannot be detected by humans.

Diffraction: "The bending and spreading of light waves as they pass through a small slit or opening. . . . When we study the diffraction of sunlight using a prism, we see a spectrum (or rainbow) of colours."[31] The first sentence is correct. The second sentence describes *refraction* not diffraction.

The final segment of the flight section is the Goal 3 objectives, namely, *Understanding Basic Concepts*. Included here are the expectations that students will "identify common applications of the properties of air, such as its compressibility and insulating properties (e.g. home insulation, tires, sleeping bags, layered clothing)."[32] That insulation has nothing to do with flight appears to have escaped the attention of the curriculum specialists.

As I grappled with this scientifically illiterate document, becoming increasingly exasperated in the process, I thought back to my recent high school visit. Talking with the students as they conducted their experiments, I had the impression they were following a recipe to satisfy some course requirement, rather than building upon a solid foundation in science. Unaware back then of the curriculum that they and their teachers were obliged to follow, I did not make much of my observation at the time; now it began to make sense.

There was no more time for fuming over the curriculum—I had a talk to prepare and experiments to devise. But how should I handle the curriculum issue? Perhaps my contact at the Science Teachers' Association would have some ideas. When we talked on the phone I sensed that he thought it best to side-step the issue. I could not do that, but how would the young teachers in the audience react when I started criticizing the curriculum? Some may have been taught under such a curriculum when they were at school themselves. The solution was obvious. First, I would captivate them with an impressive array of experiments that they could repeat in their classrooms using simple everyday items. Once I had them on my side, I would target the curriculum.

I would have to come up with some really impressive experiments, but that was no problem for someone who has written books on how to build dinosaurs from chicken bones. Being resourceful was one of the attributes one acquired growing up in post-war Britain. We youngsters had to make our own entertainment, and I was always building and experimenting with something.

CHAPTER 2

THE YOUNG SCIENTIST

Beckenham Hospital, nestled beside our local recreation ground, still had its white roof camouflaged with swirls of khaki and green several years after the war ended. I used to wonder how those patches of paint could have fooled German bomber crews into thinking the building was part of the surrounding greenery. But the hospital was never hit, so it must have worked. Many other buildings in the neighborhood were not so fortunate. A house behind our home was demolished by a bomb. Our parents attributed our escape from the blast to the poplar trees that stood in the alleyway at the bottom of our back garden.

My earliest recollection of the war was standing in the garden watching the telltale red streak in the sky of a V1 flying bomb. The last of these bombs fell on the London area in the summer of 1944, so I was no older than two at the time. My brother, seven years my senior, questions the veracity of my memory, arguing that I was merely recalling stories I had heard from him and my older sister. However, I have another wartime memory that could only have been acquired from personal experience. I vividly remember being carried over my mother's shoulder as she hurried past some men who were dismantling an unexploded bomb by the roadside. I recall counting her footsteps as she hurried on. If only I could reach one hundred, I thought, we would be safe, though I am not sure I could count that high because I was no more than three. Such details of traumatic events become etched into our minds and we now understand why.[1] The release of adrenaline into the blood—part of the fight or flight mechanism—enhances our long-term memory, helping us to avoid similar situations in the future.

Life in post-war England must have been hard on our parents. With peacetime industries in disorder and wartime debts to repay, the economy was in ruins. Money was tight—not that there was much to buy in the shops anyway with food rationing and all the other shortages. As far as we children were concerned, though, everything was fine, and I have the happiest memories of my early childhood. Toys, especially new ones, were scarce, but we made our own amusements, and often our own toys.

Most mothers sewed, and empty cotton reels—the wooden spools on which the thread was supplied—were in great demand in most households. We boys made all-terrain tanks from our cotton reels, while the girls modified theirs into devices for French knitting. All the girls had to do was hammer four small nails around the central hole at one end of the reel. This formed a square frame at the top of the French knitter, around which the wool could be looped to weave a cord of wool. Girls would spend hours weaving multicolored cords by swapping one ball of wool for another. These cords could then be coiled and sewn together into table mats.

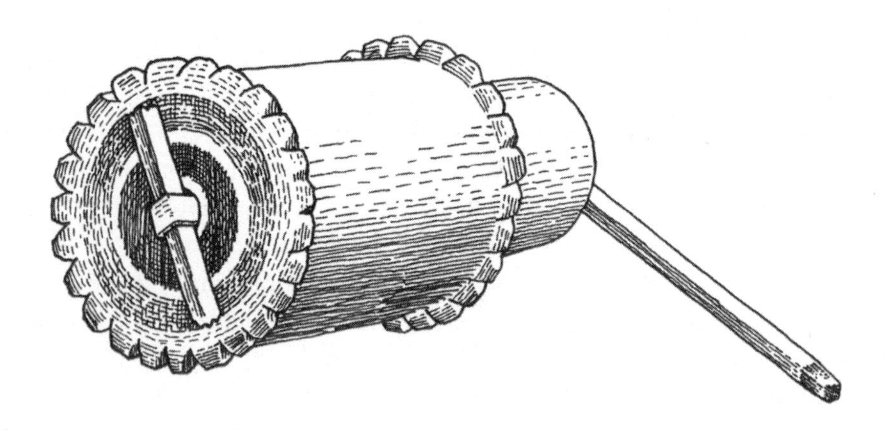

Figure 2.1: A toy tank, made from a cotton reel.

I think us boys had more fun with our cotton reels. Using a sharp knife, we would start by cutting out a series of small triangles around the edges of the reel, transforming the smooth rims into jagged ones. This was to give our toy tank better traction over rough terrain. Next, we would cut a slice from the bottom of a candle—about as thick as a little finger—taking care to make it of uniform thickness. After removing any remnants of the wick, we would widen the central hole in the candle slice, using a pointed matchstick. Once the hole was large enough, we would thread an elastic band through, snagging it over the end of a match. Holding the match in place against the candle, the elastic band was fed through the central hole in the reel. Once it poked out from the other side, it was held in place with a small piece of matchstick, too short to reach to the edges of the cotton reel. The matchstick at the candle end was then adjusted so that one end extended well beyond the edge of the reel, thereby keeping it in contact with the floor when the tank was set down in readiness for moving off.

All that remained was to wind up the tank by spinning the matchstick with one finger. Once released onto the floor or table, the tank would move along fairly slowly, due to the friction between the matchstick and candle. We used to place obstacles in its path—pencils, books, lead soldiers, our hands—and watch as it clambered over them.

Just recently I built a cotton-reel tank with my seven-year-old grandson. He played with it for a couple of hours, forgetting all about his iPod.

I was not a boy who stayed still for long, so I usually built my tanks for speed rather than for climbing. To this end I eliminated the candle. I also substituted a large darning needle for the matchstick, which reduced the friction still further. The tank would zip across the floor, beating all the clockwork cars.

Toys were usually only purchased at birthdays and Christmas, but something caught my eye in our local Woolworths store one morning, when I was out shopping with my mother. And Mr. Bob the diver was so inexpensive that she bought it for me.

Beneath the make-believe helmet and bright paint, Mr. Bob was an empty tube, about three inches long, weighted at the bottom so that it floated vertically when released into a bottle of water. According to the instructions, after topping up the bottle and inserting a cork, Mr. Bob could be made to sink to the bottom by firmly pressing down the cork. Releasing the pressure would make him rise again. With a little practice, you could make him float at any level. Mr. Bob was the source of fun for some time to come. He was also the source of puzzlement: how did it work? I do not remember how long it took me to work this out, but it would have taken far less time if the tube had not been painted. That way I would have seen that increasing the pressure inside the bottle compressed the air inside the diver, causing water to rise up within the tube to make it sink. Prior to this, the diver was neutrally buoyant, neither sinking down nor floating up. This simple device is known as a Cartesian diver, named after René Descartes (1596–1650), the French philosopher who is believed to have invented it.

Mr. Bob responded more readily to pressing down on the cork when there was no air space above the surface of the water. This is because water, unlike air, is essentially incompressible so the pressure exerted on the cork is transmitted directly to the air inside the diver, rather than being partially wasted in compressing the air in the air space. The manufacturers did not realize that their toy could be used to show children the difference in compressibility between air and water.

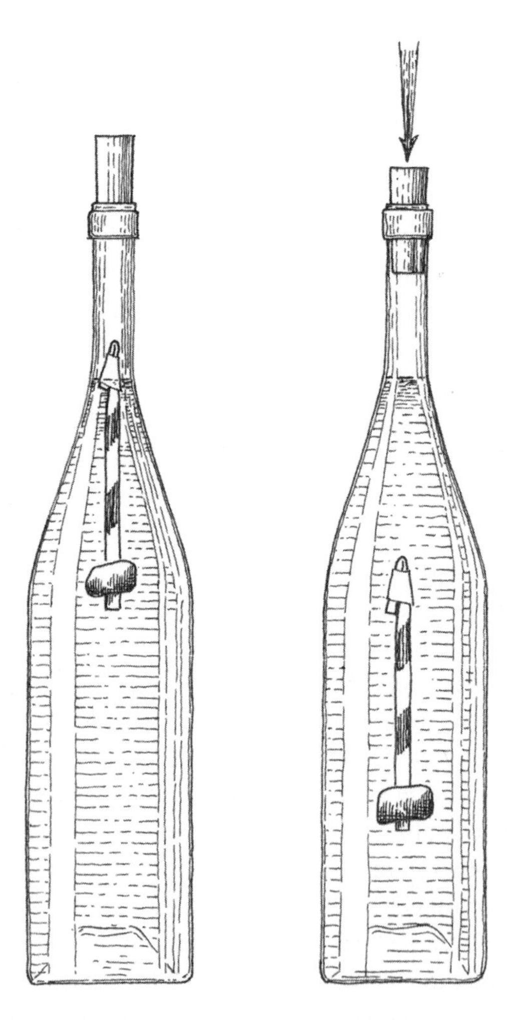

Figure 2.2: A Cartesian diver, made from a short length of drinking straw, bent over at the top and weighted with a ring of modeling clay at the bottom to make it neutrally buoyant.

Long after Mr. Bob had been broken or lost, I devised a simple substitute: a three-inch length of drinking straw, bent over and taped down at the top, and weighted at the bottom with a ring of modeling clay (Plasticine).

Exploration of the depths appealed to my boyish sense of adventure, and deep sea divers were front-page news in the spring of 1951. This was because of the disappearance of HMS *Affray*, a British submarine that was lost while on routine exercises in the English Channel.

Little went to waste back in the fifties, not even the foil caps from milk bottles. Made from aluminium, somewhat thinner than the kind used today in pie plates, they were about an inch in diameter, with rims about one-quarter of an inch deep. Free milk was supplied to schools, and this was glugged down during morning break. We would then race outside into the playground and launch our bottle tops into the air like flying saucers. The trick was to hold the cap by the rim, using the tips of your index and middle fingers. Then, by giving it a flick as you threw your hand forward, you sent it spinning on its way. The spin was to impart stability, like a gyroscope—the same stability provided by the wheels of a bicycle.

We used to see whose flying saucer could fly the farthest. The caps were essentially flat, and I discovered that doming the top, by rubbing and stretching the underside with a thumbnail, increased the range. I named my modification *The Dome of Discovery*, after one of the landmark buildings at London's Festival of Britain. This exhibition, opened by the king in the spring of 1951, was planned to help the country out of its post-war doldrums. The Dome of Discovery was where the discoveries in science and exploration were being displayed.

I did not realize it at the time, but my domed flying saucer—a miniature version of the Frisbee to come—functioned as an airfoil. In this regard, it was like the wings of the balsa wood aircraft that my brother and I used to make. Building and flying those aircraft taught me a great deal about aircraft design and the mechanics of flight, without my even realizing it.

The kits comprised thin sheets of balsa wood, printed with all the struts and cross-sectional forms needed to construct the airframe. After cutting these out and gluing them together, the finished airframe was ready for its outer covering of tissue paper. Secured with dabs of glue, the tissue could then be coated with a thin varnish, called dope, for additional strength. All that remained was to attach the motor—a thick loop of elastic fitted inside the fuselage, secured at the tail end, and connected to the propeller at the other.

I experimented with letting the aircraft take off from the ground but soon realized that this was far too uneven, especially for such a lightly constructed machine. Hand-launching was the only option, and a good push-off helped it become airborne. Flying into trees and other obstacles was a constant hazard, and much time was spent in repairs.

During a recent UK trip I visited several toy shops, hoping to buy some balsa wood aircraft kits for my grandsons, but without success. I assumed from this that such kits were a thing of the past but have since discovered that they are still available online. Could building model aircraft that youngsters can fly themselves tempt them away from their electronic devices?

English houses were notoriously cold in winter, and we would awake on chilly mornings, windows frosted with ice ferns, to see our breath billowing like smoke. There were fireplaces in two of our upstairs bedrooms, but these were never lit, and it was usually only on weekends that a fire would grace the grate in the living room. However, the iron stove in the kitchen supplied our hot water and was alight most of the time.

Coal was the fuel of the early fifties, and blue smoke curling up from home chimneys was a familiar sight on cold winter days. Unlike the warm comforting smell of logs crackling in a fireplace, this smoke produced an acrid bite that caught at the back of the throat. The same smoke belched from power plants and factory chimneys, and with all this smoke came the fog. This was not the white nothingness of a North Sea fog but a dense blanket of murk, tinged yellow, with the tang of sulfur.

Those who have never experienced a thick London fog would find it hard to believe what it was like. Imagine walking down a daytime street unable to see anything more that the silhouettes of the buildings on the other side of the road. You are alone, or so you think, but you hear muffled footsteps. The darkening shadow that appears ahead in the gloom morphs into a person. Side-stepping to avoid collision, each then disappears into their own world of solitude. While surreal for strangers living far from cities, this was a normal part of winter for Londoners. This was the fog of Jack the Ripper and of tales told by Dickens and Conan Doyle. It was also the fog that had a bearing on Darwin's theory of evolution, as I would later come to learn.

My home town of Beckenham was only a half-hour train ride from central London, and our parents sometimes took us on day trips to the capital. Trafalgar Square was a popular venue, with its fountains and statues and pigeons, and with Nelson's Column. Every schoolboy knew about

Admiral Nelson, standing atop his towering column, but I knew nothing about the stone from which his memorial was made. It was certainly black, like the column, and may have been chiseled from black marble.

The fog, dubbed *smog* in reference to the smoke, left its mark on handkerchiefs used to clear congested nasal passages. Black particles were also deposited inside lungs, and many people suffered serious pulmonary problems. The great London smog in early December 1952 claimed the lives of more than 4,000 people from respiratory disorders. This eventually led to the Clean Air Act of 1956, prohibiting the burning of coal in large parts of many urban areas. The only fuel allowed in these zones was smokeless fuel, like coke, made by baking coal at high temperatures to drive off volatile substances like coal-tar.

When we stopped burning coal, it was surprising how quickly the smog became a thing of the past and how our world began to change. Cleaning crews set to work on public buildings and monuments and, in 1968, Nelson's column was cleaned. I was amazed to discover that Nelson's statue was made from sandstone, of a similar light color as the granite column. Changes began to appear in the living world too, some of which relate to Darwin's theory of evolution by natural selection.

Natural selection was Darwin's way to explain how evolution came about. He reasoned that each species produces far more progeny than can survive. Offspring are not exactly alike, and some have features that give them advantages over others. This gives them better chances of survival, so they leave more offspring. Since these offspring inherit some of their parents' favorable features, they too have improved chances of survival. The action of natural selection, operating over long periods of time, eventually leads to the appearance of new species.

One of the classic examples of natural selection in action, described in most school textbooks dealing with evolution, is the story of the peppered moth, *Biston betularia.* The peppered moth has two common forms, or morphs, a black one and a speckled gray one. During the Industrial Revolution, trees and other plants became darkened with soot, along with buildings. Naturalists noticed that the black or melanic morph was more numerous in polluted areas than the more common gray morph. This was attributed to the melanic forms being better camouflaged against the darkened trees than the others, making them less vulnerable to hungry birds, and was referred to as industrial melanism. These observations were being made toward the end of the nineteenth century, but it was not until the 1950s, and the work of Bernard Kettlewell of Oxford University, that the findings became widely known. The story of the peppered moth was hailed as compelling evidence for evolution through natural selection.[2]

During the last several decades many criticisms have been made of Kettlewell's work, casting doubts on its validity. Creationists have seized upon this as the demise of Darwin's theory, but they overlook the facts. Kettlewell's principle experiments were completed before the beneficial effects of Britain's Clean Air Act became apparent and were conducted over a relatively short time period. Since then, several long-term studies have been conducted that document changes far more dramatic than those described by Kettlewell. In one study, involving more than 18,000 peppered moths, collected over a period of more than thirty years, the melanic morph declined from above 90 percent of the population to below 10 percent. These changes paralleled declines in soot and other atmospheric pollutants in the industrial area in which the study was conducted.

Growing up during the London smogs gave me insights into industrial melanism that would last a lifetime. Whenever I think of the peppered moth story, I visualize Nelson's column, before and after the cleanup.

Rabbit stew and dumplings—of all our mother's meals, this was the family favorite. My interests in the dish had as much to do with what we left on our plates as it did with the food itself: I collected the bones. My interest in bones was nurtured by our father. A baker by profession, he would have been a doctor if his parents could have afforded to keep him at school. He channeled his medical interests into practicing first aid with the Civil Defense. This voluntary organization was established soon after the war, primarily for the perceived threat of nuclear war with Russia. There were several medical tomes in his bookcase, and I was attracted to the sections on the human skeleton. By the age of nine I could name all the bones.

One day our form teacher at school (junior school in the UK) asked each of us about our hobbies. There were all the usual pastimes: collecting stamps, French knitting, playing with puppets, Meccano sets, train spotting . . . then it was my turn. "I study the bones of people's bodies," I said, almost apologetically. The other pupils reacted with giggles or confused silence, adding to my discomfort. Our teacher looked at me quizzically. I was by no means one of her star students, and my unusual confession must have taken her aback.

"What's this onc called?" she asked, tapping her upper arm.

"The humerus," I replied without hesitation.

"And this?" She ran a finger along her collar bone, and I replied, "The clavicle." I correctly identified all the other bones she queried, but I do not recall her being particularly impressed. This was the same teacher who had scorned my failed attempt to swim a width at the local swimming pool. Her remarkably inapt poolside comment, "You poor fish," did nothing to help my lack of self-confidence. Teachers, good and bad, can have such profound effects on their students.

While I was at that school, I received a very special Christmas present from my parents: my first microscope. Unlike many of the plastic toys sold today as microscopes, this one was a scaled-down version of the real thing, with a metal body and good optics. It was also supplied with glass slides and a dropping pipette, so youngsters could examine droplets of water for microscopic life.

We had a pond in our backyard (garden in the UK), and I used to collect water samples and search for *Amoeba*; I had an illustrated biology book and learned to recognize this common single-celled animal. I also collected small samples of the filamentous blanket weed that floated in large patches on the surface of the water. By teasing this out on the slide, I could see the individual plant cells that were joined, end-to-end, forming the long green filaments. Looking down the microscope at these tiny patches of *Spirogyra*, as I learned the plant was called, was like peering through a dense jungle, with strange animals gliding here and darting there. Many hours were spent completely absorbed in the mystical world beyond the microscope.

Like many youngsters, I used to go out with a net and jam jar in the spring to collect frog spawn from local ponds. The jelly-like clumps of transparent eggs were easy to spot floating on the surface because each spherical egg had a jet black dot in the middle. After taking the spawn home, we would put it into an old goldfish bowl filled with water. By keeping watch for the next five or six weeks, we would be able to see how the black dots grew and developed into tiny tadpoles. The remarkable opportunity of watching an embryo developing inside an egg, unhampered by any concealing shell, was taken completely for granted.

Once the tadpoles hatched, we would watch the fish-like creatures—all head and tail—undergo their remarkable transformation into four-legged animals that could walk on land. We did not keep the small frogs for long; our father did not want them in the garden pond so we would put them into jars and take them back to where we found the spawn.

Much of my last year at the junior school was spent preparing for the dreaded eleven-plus examination. This was the nationwide assessment for eleven-year-olds to determine who would succeed and move on to the local grammar school. There, they would receive an academic education, with the prospect of going on to university. Failure relegated pupils to a secondary modern school to receive a generalized schooling, with little prospect of continuing on to secondary education. In preparation for the eleven-plus exam, we were given practice questions to answer in mathematics, English, and in general comprehension. There was also the intelligence-test, with questions like: *Woman* is to *crowd* as *drop* is to *fall, flock* or *water*. Similar association questions were posed using stick-diagrams instead of words. One question that stays in my mind was in the English section. By using appropriate punctuation, sense had to be made of the following: *Smith in an examination had had had while Jones had had had had had had had had had had the approval of the examiners.* Even when properly punctuated, the solution is still poor English: Smith, in an examination, had had "had" while Jones had had "had had." "Had had" had had the approval of the examiners.

My teacher probably expected me to fail, but not my mother, who doted on her youngest child. I did not pass, nor did I fail outright. Instead, I was invited to attend for an interview at the local grammar school for boys; the final decision rested upon my performance. I don't remember much about the interview. There was a waiting room, where some other nervous boys in short trousers were seated. My name was duly called, and I crept into the presence of the omnipotent headmaster. He asked the usual questions: sports, hobbies, favorite subjects, father's occupation, what I wanted to be when I grew up. Then he asked me to read a passage from a book.

"No news is good news," cheered mother as we waited for the final result. She was still as confident as ever, and even I was beginning to think I might stand a chance. Then the letter arrived. Failure. Mention was made of my poor diction. My mother was devastated. Her youngest son, the little boy who loved animals and who wanted to be a vet, was not going to the grammar school. I was disappointed too, but not nearly to the same extent.

Gordon Sumner, the outstandingly accomplished Sting, passed his eleven-plus. In his autobiography, he railed against the "institutionalized cruelty" of separating children into academic and non-academic streams. His best friend, Tommy Thompson, failed his scholarship, and Sting had to witness his dejection at having to go to that *other* school. The two friends of almost six years—half their lifetimes—inevitably drifted apart.[3]

Despite my disappointment in failing at such a susceptible age, I believe streaming is a good practice, provided there is some safety net so late developers can subsequently switch schools, or classes. The alternative strategy to streaming—exemplified by the UK's comprehensive school system and by schools throughout North America—is to mix pupils of differing abilities. The rationale for mixing smart with slow, disruptive with attentive, engaged with disinterested, is that the better pupils will have a positive effect upon the others. Such an idealistic view would likely not be held by those with classroom experience.

Alexandra Secondary School for Boys occupied an unimpressive building in Penge—Beckenham's poor neighbor. The thought of attending such a school, a senior school at that, should have been a daunting prospect to be dreaded throughout the summer holiday, but I have no such recollections. Any anxieties I might have had were soon dispelled at the start of term: Alexandra turned out to be a fine school. This was primarily due to the teachers, teachers who took a genuine interest in their pupils. They knew their subjects and how to teach, inspiring and challenging their young charges.

Mr. Rees, a small Welshman, taught history, bringing the subject alive with his impromptu histrionics. And when he sang *Jerusalem* at the Christmas concert that year, we boys were as moved as our parents. English was taught by a whimsical character named Mr. Biggar, who also taught music. He gave us a solid grounding in English grammar, spicing the subject with his animated readings from books. One of the stories that seized our imagination was *The Man-Eating Leopard of Rubraprayag*, by the early-twentieth-century big-game hunter Jim Corbett. It was an excellent choice for boys of our age. The book, first published in 1948, is still available in paperback. Our form teacher was a tall young man named Hugh Jordan, who taught science and mathematics.

Alexandra School boosted my self-confidence and inspired me to learn, transforming me from a virtual dunce and eleven-plus failure to the top of my class. I came first in English and close to the top in science and mathematics. I even got a B+ for physical education! Already interested in science, my home experimentations took flight.

I dabbled in building rockets. For ignition I used a short length of resistance wire connected, by a long cable, to my electric train transformer, located a safe distance away. All too often, though, my rockets blew up rather than lifting off. This was hardly surprising, as I used a mixture of saltpeter (potassium nitrate), sulfur, and powdered charcoal as the propellant: gunpowder. One dark night, I attempted to launch a rocket improvised from a brass thermometer case. On activating the ignition circuit, there was an enormous explosion, followed by a deathly silence. Then all the dogs in the neighborhood began barking. My mother, fearing the worst, could not bring herself to go out into the garden. When I stepped indoors, ears still ringing, the relief on her face was palpable; she was convinced I had blown myself up.

Not all of my experiments involved explosives. I recall investigating the electrical conductivity of soil by driving metal rods into the ground and connecting them to my train transformer, using a flashlight bulb as an indicator of electrical flow. I also experimented with various devices, like a gadget for pulling back the bedclothes when the alarm clock went off.

The bedcover contrivance, built to see if I could accomplish the task rather than for practical purposes, was simple and unsophisticated. My alarm clock had a metal tab at the back for winding up the bell, and this rotated when it went off. After firmly securing the alarm clock to the furniture beside the window, I took a length of string, knotted one end, and wedged this beneath the metal tab to anchor it. When the alarm went off, this would release the string. The free end of the string was then looped over the end of the curtain rod. Filling a bag with books, I tied this to the string, as high off the ground as possible.

Taking a second length of string, I tied one end to the top of my sheet, just below the pillow. Leading the other end across the room, I looped it over the curtain rod and tied it to the bag. All was now ready. When the alarm went off, the bag fell to the ground, pulling the bed-sheet string behind it, along with the covers.

Far more elaborate than this was the helicopter device I built to raise money for the school fair that summer. The idea was to have a helicopter, mounted at one end of a long plank, which would fly around in circles a few feet above the ground, and release a metal "bomb" into a bucket, winning the player a prize. The helicopter was simply an electric motor fitted with a large model-aircraft propeller. The motor, which ran off the household power supply, was the one that my father used to power the water pump for the fountain in our pond. The bomb-release was a battery-powered electromagnet that I built myself. Everything was simple enough—but how could I supply electricity to the rotating system?

My solution was to make two pairs of metal collars cut from tin cans—one pair for the electric motor, the other pair for the electromagnet. Each metal strip, separated from its neighbor by a gap, was wrapped around a wooden drum that was attached to the underside of the plank, at its midpoint. I have long since forgotten what I used for the drum, but this sat on top of the sturdy turntable that my father used when icing cakes. As the plank spun around, so did the drum beneath, with its four encircling metal bands. A pair of wires connected each double band with its respective device. Electricity was supplied to each pair of rotating collars by a pair of springy metal tabs, pressed firmly against the collars to maintain close contact during rotation. Modern health and safety concerns would never permit such a contrivance on school property today, but it was no problem back in the fifties. The helicopter worked so well that some of the parents borrowed it for a fete at another school.

With Britain's peacetime manufacturing in shambles, the brightest light in the sky was the de Havilland Comet, the world's first jet airliner. Looking elegantly sleek with its four jet engines buried in the wings, the aircraft was quieter and faster than the most advanced piston-engined airliners. It could also climb faster and higher, allowing it to fly above bad weather conditions that rivals had to contend with. I would have proudly hung a model Comet beside my plastic Spitfire if I could have found one in my local toy shop.

In addition to domestic sales to BOAC (British Overseas Airways Corporation, now called British Airways), Comets had been sold to France and were on order to the United States, Venezuela, Brazil, India, and Japan. Delighted passengers commented on its quiet, smooth flight, and de Havilland could boast that the Queen and other royals had flown aboard the Comet. The future of the British aircraft industry could not have looked more promising.

On January 10, 1954, BOAC Flight 781 took off from Rome's international airport bound for London. Twenty minutes later, at an altitude of approximately 27,000 feet, the Comet broke up, crashing into the Mediterranean off the island of Elba. Everyone aboard was killed.[4] This was the third fatal crash of a Comet. The first, which had been attributed to pilot error, had occurred the previous March, in Karachi, Pakistan, when the aircraft failed to clear the runway. Two months later a Comet crashed during a severe thunderstorm, soon after taking off from Calcutta, India. That crash, caused by the structural failure of the wings, was believed to have been due to the pilot inadvertently overstressing the aircraft while pulling out of a steep dive.[5]

Immediately prior to the loss of Flight 781, her captain made contact with a pilot aboard another BOAC aircraft. He was about to ask that pilot a question, but the transmission abruptly ended in mid-sentence.[6] Witnesses from Elba reported that the aircraft had disintegrated, and some saw parts falling into the sea in flames. Local authorities were notified, and boats were dispatched to the scene. Floating wreckage was recovered, along with fifteen bodies. Autopsies revealed that the victims had died from heavy impacts with the aircraft.[7] Before year's end, the Comet crashes would enter into my world of experimentation.

On receiving news of the accident, BOAC immediately suspended all Comet flights so that detailed inspections of the aircraft could be made. This was done in collaboration with de Havilland and the civil aviation authorities. A committee was set up to investigate what modifications were necessary before the Comet could resume passenger services. Meanwhile, the Royal Navy was mounting a salvage operation to recover wreckage from the seabed off Elba.

According to the investigating committee, there were several possible causes for the accident. These included problems with the flight controls, an engine fire, defects in the windows causing them to break, triggering an explosive decompression, and metal fatigue.[8]

Metal fatigue is where a metal is weakened, over a period of time, by repeated exposure to stresses that would normally be too small to break it. You can simulate metal fatigue with a simple experiment. If you opened out a paperclip by bending the two loops apart, it would likely not break because the stresses are too small. However, if you repeatedly bent it back and forth, the metal would fatigue and the clip would break. Metal fatigue is cumulative, and the longer a metal structure is in use the more likely it is to fail. Fortunately, fatigue failures are rare—consider how many millions of times the connecting rods of a car engine move up and down, but they rarely ever break. If things are properly built and maintained fatigue is not an issue.

When all the inspections and tests had been made and the necessary modifications completed, the committee concluded that "everything humanly possible has been done" to ensure safety and that the Comet should return to normal operations. The entire process had taken just ten weeks.[9]

Flights resumed on March 23, 1954. Sixteen days later, a Comet crashed into the Mediterranean near Naples, under circumstances similar to those of the Elba disaster. The Comet fleet was immediately grounded, and its Certificate of Airworthiness revoked.[10]

What followed was one of the most remarkable pieces of detective work in the history of aviation. This was conducted at the Royal Aircraft Establishment at Farnborough, in the UK county of Hampshire. The prime suspect in the two Mediterranean disasters was metal fatigue, caused by repeated pressure changes in the fuselage as the aircraft climbed to its cruising altitude of 40,000 feet and then descended to land. At 40,000 feet, the outside pressure is only one-fifth that at sea level, causing the fuselage to expand slightly. The reverse happens on descent, causing the outer skin to contract. The recurring changes in pressure during flying operations, akin to inflating and deflating a balloon, repeatedly stressed the fuselage, leading to metal fatigue.

The investigators had a Comet at their disposal so they could test this hypothesis experimentally. A watertight testing tank was built around the fuselage so it could be immersed in water. Both the tank and fuselage were then flooded. The fuselage was now pressurized by pumping in more water, continuing until the pressure difference, inside and out, was comparable to that of an aircraft at cruising altitude. The extra water was then released, so the inside and outside pressures were the same again, simulating one complete takeoff and landing cycle. By repeating the process around the clock, the investigators could accumulate hundreds of simulated flights. The reason for using water rather than air in the experiment was to avoid extensive damage to the fuselage in the event of its rupturing. If air was used and the skin split, the fuselage would explode, losing valuable information. This would not happen with water because, being essentially incompressible, there would be no explosive release of pressure as the fuselage failed.[11]

The Comet in the testing tank had made 1,230 flights prior to the start of the experiment. Testing began in early June and continued until June 24. That was when the fuselage failed, after 1,830 simulated flights, for a total of 3,060 real and replicated takeoffs and landings. The failure started at the corners of one of the cabin windows.[12]

By this time, much of the wreckage of the Elba crash had been recovered, most of which had been reassembled like a giant jigsaw puzzle. Examination of the top of the fuselage showed a large split that passed through the corners of a pair of windows used for the direction-finding aerials. The investigators concluded that this was where the first fracture of the cabin had started. As engineers know, square-cornered, as opposed to rounded-cornered openings in structures under load, concentrate stresses, thereby increasing the chances of initiating a crack. Such cracks rapidly grow, ultimately leading to structural failure. Here was confirmatory evidence for the cause of the Elba crash. As the fuselage blew apart during the explosive decompression, people were thrown into the air, sustaining the fatal injuries discovered during the autopsies.[13]

Hearing the news that the Comet crashes had been caused by metal fatigue fired my imagination. I wanted to see if I could fatigue a piece of metal to its breaking point, but I could not devise a suitable experiment. Then I had an idea. We had a swing in our back yard. What if I tied a bottle to a piece of string so I could drag it back and forth along the path beneath the swing? This would expose the glass to forces that were too small to break it, but the accumulated effect might cause it to break due to fatigue.

I do not remember for how long I swung back and forth on the swing, but it was probably for several minutes. I began to wonder whether anything would ever happen and whether I should abandon the experiment. Then, without any warning, the bottle shattered. Glass fatigue!

Looking back in wisdom many years later, I rejected the idea that fatigue had broken the bottle. Fatigue was surely a property of metal, not of glass—the bottle must have just clipped something on the path.

Does fatigue failure occur in glass? On recently reviewing the literature, I found some disagreement. According to some specialists, glass does not fatigue like metal. Others argue that glass *does* fatigue and that this was not generally recognized until the late 1970s.[14] Maybe my 1950s experiment was ahead of its time.

Modifications to the Comet included replacing square windows with oval ones and strengthening the skin of the fuselage. Commercial flights resumed in 1958, but competition from the Boeing 707 and the Douglas DC 8 was so fierce that Britain lost her lead in the sky to America.[15]

I do not remember sitting a thirteen-plus exam or facing any interview but, in the summer of 1955, I was transferred to Beckenham Technical School. Technical schools were essentially grammar schools with more applied subjects, like technical drawing and metalwork, and fewer academic ones, like Latin. I was now competing with more capable students, and my marks and class standings declined accordingly. But things improved over the next couple of years, except in subjects that were unimportant to me, like woodwork.

At the start of my second year at Beckenham, science was split into physics and chemistry. I immediately took to chemistry, soaring to the top of the class and staying there for almost the entire time I studied the subject. I devoured the chemistry textbook like a novel, conducting experiments at home as well as at school.

There was a certain drugstore in our area, well known to experimentalists like myself for selling us anything we wanted. The shop was usually crowded with boys on Saturday mornings, buying everything from concentrated acids to the components of explosives. How the pharmacist escaped prosecution for such reckless endangerment escapes me. I used to buy various chemicals, including concentrated acids, along with test-tubes, so that I could repeat some of the experiments that our chemistry teacher had demonstrated for us in class.

School took a more serious turn in the fall of 1957 when we fifteen-year-olds entered the fifth form, equivalent to grade ten in the United States. This was when we began preparing for our O-level GCE exams in various subjects, which we were scheduled to take the following summer. The Ordinary-level General Certificate of Education, the nationwide assessment of academic achievements, followed the same syllabuses across the country. Exams were set and marked by external examination boards, the two most widely used ones being those of the University of London and the University of Oxford. The examination papers and marking standards were similar from one board to another, so it made little difference which board was selected.

The GCE exams were searching and challenging and, since answers had to be reasoned and clearly written, achieving good grades required more than recalling relevant facts. Students aspiring to attend university took at least five O-level subjects, which included English and mathematics, along with their subjects of choice.

Many in our class left school at sixteen, but most of my close friends stayed on into the sixth-form for two years—equivalent to grades twelve and thirteen—working toward their A-levels. The advanced-level GCE examinations required greater depths of understanding than O-levels, and most of the science subjects included a practical exam. The minimum university entrance requirements were five O-levels and two Advanced-levels, and some universities had additional requirements. For example, prior to 1960 students would not be accepted at Oxford without having passed O-level Latin.

My school did not offer A-level chemistry, nor biology (even at O-level), so the only science I took in the sixth form was physics. And it was in physics, where we spent most of our time conducting experiments, that I received my first training as a scientist. Our teacher, Mr. Faires, was a cynical character with a pragmatic and inquiring mind. He encouraged us to ask questions, to think analytically, and never to take anything at face value. With his focus on understanding basic concepts and learning through inquiry, he infused me with the very essence of science. If there were more people like Ron Faires in the world, there would be less gullibility to the pseudoscientific nonsense to which we are exposed in these anti-intellectual times.

CHAPTER 3
IGNORANCE IS BLISS

Some of the most flagrant examples of scientific illiteracy are provided by the health, wellness, and cosmetic industries. Take, for example, the absurdity of some of the shampoo formulations advertised by manufacturers. Most shampoos contain vitamins, proteins, and other nourishing ingredients, but how can hair, a non-living material formed of the same protein (keratin) as our nails and the outer layer of our skin, consume *any* nutrients? There are also hand creams containing vitamins, wheat germ, honey, kiwi fruit, and other tasty things, along with nail polish enriched with proteins and vitamins.

And what about all those healthcare devices on the market offering countless cures and benefits, none supported by any credible evidence? Magnetic bracelets are claimed to alleviate everything from arthritis to epilepsy, while, for the more mechanically minded, there is an innovative foot rocker that revitalizes swollen and aching legs, restoring blood circulation. Why exert yourself walking when you can get all the same benefits sitting and relaxing in the comfort of home. My all-time favorite is the detoxifying footbath, where users can sit and watch as the harmful toxins are drawn from the body, discoloring the surrounding water. No need for a liver when you have this device to purify your blood! The device is, of course, a complete scam.

While many people make use of remedies and devices in their homes, others seek the help of the various clinics offering alternative medicines. The last few years have seen a proliferation of such clinics, ranging from acupuncture and homeopathy to reiki and spinal decompression. The most familiar among these alternative practices—one that has been broadly used and accepted in our western world for generations—is that of the chiropractor.

People seeking the help of chiropractors often talk of having their vertebrae put back into their proper place by manipulation, as if the bones were free to move. However, the vertebrae are so securely held together that they can remain in alignment even after a person has been involved in a motor accident severe enough to break bones. The integrity of the vertebral column is largely attributed to the intervertebral discs. These multifunctional structures hold adjacent vertebrae together securely with a sufficient degree of movement to give the vertebral column its flexibility. The discs also serve as spacers, preventing bone-to-bone contact, and as shock-absorbers, reducing the stresses of impacts. I discovered the remarkable binding abilities of the intervertebral discs when I dissected Tantor, a six-ton elephant that died at the Toronto zoo in 1989.

By a long-standing arrangement with the zoo, the Royal Ontario Museum has the option of obtaining carcasses of animals that die there, so they can be skeletonized and added to the ROM's osteological collections. Tantor died following surgery for a chronic tusk infection, and the body was transported to the Ontario Veterinary College, at the University of Guelph, where a postmortem was conducted.

As it happens, a long weekend intervened and, by the time our small group from the museum started work, the elephant had been dead for six days. The first thing that struck me as we began removing the thick hide and hacking flesh from bone was that the body was still warm. Here was a graphic example of the relationship between areas and volumes. Heat is lost from surfaces and, as things get bigger, the ratio of surface area to volume rapidly diminishes. This explains why the peas on your plate cool down much faster than the baked potatoes.

The animal was lying on the floor, on its left side, and we began by removing both of the right legs, along with most of the hide. Fortunately, we had an overhead winch to assist with this monumental task. We then decided to sever the vertebral column, just in front of the hip region. Before we could do this, we had to remove the overlying muscle. Eventually the vertebral column was exposed and we winched up the back end of the carcass. This placed a considerable tension on the vertebral column to assist with the separation of the vertebrae.

Wielding the sharp carving knife that I had used for removing hundreds of pounds of prime steak (unfortunately this could not be eaten because of all the drugs) I began separating the ligaments and small muscles holding the two targeted lumbar vertebrae (lower back region) together. This caused the vertebral column to yield a little, but it remained intact. All that was holding the two vertebrae together now was the intervertebral disc, together with the ligaments and muscles of the other side.

Trying to find the position of the intervertebral disc was difficult, and, every time I stabbed out at what I thought was the right place, I hit bone. Eventually the tip of the blade sunk into something soft—I had found the right spot. Instantly and spectacularly the disc tore apart and the vertebral column separated, illustrating the effectiveness of the discs in keeping the vertebrae together and in alignment. I have this mental picture in mind whenever I hear of people having their vertebrae realigned, or when seeing chiropractic images of misaligned vertebrae.

Visiting home shows is a mixed blessing for me. I enjoy sampling in the food courts and seeing some of the weird and wonderful new gadgets being promoted, but I get twitchy when my wife steers me toward any kitchen renovation booths. Our last home show was different, though, because I was on a mission. I wanted to check out the health and wellness booths, and the first one I headed for belonged to a local chiropractor.

The helpful young man with whom I spoke was a doctor of chiropractic. This was my very first encounter with someone of his profession, and I proceeded with an open mind. He began by explaining that the objective of chiropractic was to correct for kinks in the spine, abnormalities that cause pinching of the spinal nerves and bring about health problems. I was then invited to undergo a check for my posture. For this, I was asked to stand upright on a pair of scales, a foot on each one, and to look straight ahead. This was to check for differences in loading between my left and right sides. I was not surprised when he announced that there were differences in readings between the two scales, in the order of three or four kilograms (7–9 lbs.). While I was still standing on the scales, he touched my shoulders and then my hips. Then, using both hands, he felt in the small of my back.

Next, he asked me to turn my head as far as I could to the left, and then to the right. He said I had a twenty percent reduction in neck mobility. He also said that my left shoulder was lower than the right, as was my pelvis, showing that my vertebral column was kinked. When I inquired on the likely extent of the curvature, he said he would have to take X-rays to determine that.

I suspect that many people undergoing the twin-scale test would be concerned at discovering anomalies between their left and right sides.

Being told that their shoulders and hips were also tilted would seem to confirm that the backbone was kinked. But consider the facts. First, what are the chances of being able to stand on a pair of scales with your body weight distributed *exactly* equally between the two? Probably close to zero. If you have a bathroom scale, you can test this for yourself. Start off by placing a book of similar height beside the scale. Now stand with one foot on the scale and the other on the book. You will see how much the scale changes as it tries to register your half-weight. If you try changing your posture from side to side, even slightly, it sends the scale into turmoil. I would challenge anyone to stand on a pair of scales and register the exact same weight for each foot. I would also challenge anyone to register the same pair of dissimilar weights on consecutive tries. The twin-scale test is meaningless.

A second point to consider is that the left and right halves of our bodies are not exactly the same: the left half is not a mirror image of the right. This is true internally as well as externally. For example, the left side of the large intestine (colon) is longer than the right, the right liver lobe is much larger than the left, and most of us know that the heart is inclined toward the left side of the chest. Some of us are right-handed and others left-handed. The muscles on the dominant side are usually larger and stronger, and there are corresponding differences in the bones. It should therefore come as no surprise if one shoulder or hip were held slightly higher than the other.

The last and most significant point about my poor showing in the chiropractor's tests is that I am devoid of any back pain or discomfort. Like anyone else of my age, I get the occasional twinges, especially if I have been doing any heavy lifting, but this is completely normal. I would be very surprised if I had any abnormal curvature of my spine.

I discussed the "kink" in my vertebral column with the chiropractor. He said that when they see children and young people with such kinks they can be corrected. Then, to his credit, he said that at my age this was not possible and that any chiropractor who said otherwise was not doing a good job. However, proper chiropractic treatment could prevent the condition from worsening, thereby avoiding additional kinking and further loss of neck mobility. He thought my condition was probably already causing vertebra-to-vertebra contact at the edges of some of the vertebrae. Had that been the case, I am sure I would have been in considerable pain. He told me I could have a full assessment of my back if I wished. The normal cost was $180, but they were offering a show special of only $20. Naturally I declined.

A second chiropractor I visited was particularly interested in the natural curvature of the neck. This, I was informed, can deteriorate, leading to the condition known as *forward head posture*. His handout included diagrams reminiscent of images we have seen of prehistoric man, head slouched forward. The health problems allegedly caused by this condition included pinched nerves and disc deterioration.

He explained that there was a simple diagnostic test for the disorder. All one had to do was stand with the heels, pelvis, shoulders, and back of the head touching a wall. If this felt like an uncomfortable position to maintain, you had the ailment. This is confirmed when your head moves forward as you relax into your normal posture.

I took the test and, needless to say, was diagnosed as having the condition. However, I suspect most people would arrive at a similar conclusion after conducting such a test. Fortunately, the condition could be corrected, he explained, by using a head block and weights. This course of treatment would stretch the neck ligaments, thereby restoring the spine's natural curvature. The thought of lying on one's back on a chiropractic table, neck on block, head pulled back by weights, is the proverbial cure being worse than the disorder.

The next booth I visited, operated by a naturopathic and integrative health clinic, introduced me to a variety of specialists in alternative medicine. The first lady I met was a Reiki practitioner. She spoke to me about the body's internal energy, and how things can happen to interrupt its natural flow. When I asked if external sources of energy, like sunlight, were involved, she said no, Reiki all had to do with internal energy and realigning its flow. This realignment increased the body's self-healing ability, reducing stress and enhancing relaxation. What, you may ask, is this internal energy flow?

William Lee Rand, founder and president of the International Center for Reiki Training, in Michigan, and a leading Reiki practitioner, explains how

> We are alive because life force is flowing through us . . . [it] flows within the physical body though pathways. . . . It also flows around us in a field of energy called the aura. Life force nourishes the organs and cells of

> the body, supporting them in their vital functions. When this flow of life force is disrupted, it causes diminished function in one or more of the organs and tissues of the physical body. . . . Reiki heals by flowing through the affected parts of the energy field and charging them with positive energy.[1]

Finding little to discuss with the Reiki practitioner, I got into conversation with a second lady, a holistic nutritionist. As a registered practitioner of her profession, she recognized that each person is biochemically distinct, with unique nutritional requirements. Her goal was to help her clients, through guidance and education, to find the diet best tailored to their individual needs and preferences. Her philosophy was to eat whole and unprocessed foods, preferably vegetables that were raw or lightly cooked, and to limit the consumption of animal products. This made good sense, even though it had little personal appeal. But then we started talking about detoxification.

There were toxins (poisons) in the environment, she explained, which had to be removed from the body. Certain plants were useful for detoxifying, like parsley, which should be included in the diet. Periodically, however, perhaps once or twice a year, extreme detoxification is required. When I asked if this involved eating parsley for a day, I was told it had to be more than that. Among the preferred detoxifying agents are beets, avocados, lemons, flax seed, garlic, and seaweed. I asked about detoxifying foot baths, hoping to gain further insights into this fake device. Although she had nothing to say about them, she did comment about the use of Epson salts in the bathtub for detoxification. This evoked childhood memories of my mother, who periodically administered Epson salts (magnesium sulfate), or senna pod tea, to me and my unwilling siblings to "purify" our blood. Both are laxatives, which, I suppose, was their blood purifying mechanism. We learned to survive the purification ritual by pouring the awful stuff away when she wasn't looking.

The third specialist in the booth was a naturopathic doctor who was also the founder of the clinic. Many naturopaths describe themselves as doctors, some having obtained their doctorates from one of the institutions accredited by the Council of Naturopathic Medical Education in Massachusetts. While carrying a stethoscope and equating their medical training with that of a medical school graduate, they are *not* medical doctors and are unable to fulfill that role. Accordingly, naturopathic doctors are generally not licensed to carry out surgical procedures or to prescribe drugs, though there are some significant differences between individual American states and between individual Canadian provinces. Oregon, for example, allows licensed naturopaths to prescribe drugs and perform minor surgery, and British Columbia became the first province in Canada to grant prescribing rights to naturopaths, back in 2009.[2] Other practitioners in the fields of alternative medicine, including chiropractors, describe themselves as doctors, but they too should not be confused with medical doctors.

Naturopaths are among the generalists in the field of alternative medicine. According to the website of the Association of Accredited Naturopathic Medical Colleges in Washington, DC,

> Naturopathy is a traditional approach to health that is holistic, meaning that it encompasses the whole being. It is based on natural and preventative care. Naturopathic medicine combines many methodologies, such as acupuncture, massage, chiropractic adjustment, homeopathy and herbal cures, along with sensible concepts such as good nutrition, exercise and relaxation techniques.[3]

The treatments used by the doctor of naturopathy whom I met included homeopathy, lifestyle counseling, and traditional Chinese medicine, her specialty being acupuncture. I did not get the opportunity to discuss acupuncture with her but did get some insight by visiting the clinic's website:

> Traditional Chinese medicine explains acupuncture as a technique for balancing the flow of energy or life force—known as qi or chi—believed to flow through pathways (meridians) in your body. By inserting needles into specific points along these meridians, acupuncture practitioners believe that your energy flow will re-balance.[4]

I was surprised to discover that courses in traditional Chinese medicine are offered at the University of Minnesota. Remarkably, they have a Center for Spirituality & Healing, offering a range of courses, including Acupressure; Yoga: Ethics, Spirituality, and Healing; Reiki Healing; and even one titled Horse as Teacher: Intro to Nature-Based Therapeutics Equine-assisted Activities & Therapies.[5] Judging from some of the outraged comments on the Internet, I am not the only one disturbed to see the distinction between fact and fancy being blurred at an academic institution.

Having had a brief introduction to acupuncture, I visited a booth specializing in Chinese medicine. The gentleman who attended me was an acupuncturist, and he asked if I would like him to perform a pulse diagnosis, to determine what medical problems I had. After accepting his offer, I was asked to fill out a consent form and to mark off the problems I had. I ticked the boxes for sleep and for anxiety/stress. Taking the offered seat, I presented an arm, sleeve rolled-up. He explained how three things were to be detected on each side of the wrist and reeled off heart, lungs, stomach, spleen, and, I think, intestines and kidneys.

I asked, seemingly innocently, what the spleen did, and he said it produced cells to combat infection, which was correct. He then incorrectly said that it worked with the liver to assist in digestion.

After a minute or so of feeling my pulse, he said he had detected a skipped beat. This made me stop and think; I've had an irregular heartbeat for years, and my heart had been banging around all over the place at the time. Had he detected the problem? That would have been impressive.

He returned to monitoring my pulse. A few minutes later, he reached his conclusion: I had issues with my heart, liver, and spleen. Acupuncture, however, would restore things to normal. Thinking I was a likely new client, we sat and chatted about acupuncture. One of the things I asked was whether the needles hurt. He assured me that the needles were very small, likening their effect to a mosquito sting. Most people, he said, felt nothing after the first needle. Taking the proffered business card, I thanked him and departed for the next booth.

Here was something completely new to me: bioenergetics, or bioenergetic intolerance elimination (BIE) to use the full name. There were two ladies in the booth, the owner of the operation and her assistant, who was busy with a client undergoing a test. While I chatted with the owner, I noticed a wooden box on the table, packed with finger-length glass vials. Each was filled with a clear liquid, capped and identified with a neatly printed label; they might have come from a lab. Noticing my interest, she removed one of the vials, and I checked the label: serotonin, a neurotransmitter, associated with mood swings, among many other things.

She explained how certain substances can cause us to have negative reactions, resulting in symptoms like fatigue, poor memory, and digestive disorders. The long list of symptoms displayed at the back of her booth ranged from acne to migraine. The things that triggered these problems could be detected by holding a vial against the wrist area and testing the subject's muscle response to that particular agent. A weak muscle response indicated a negative reaction.

Pointing to the serotonin vial, I asked if that was what was inside. She said no, it was just a mixture of alcohol and water. This made absolutely no sense to me, so she explained how every substance has a "signature frequency" that our bodies are able to recognize. It is these signature frequencies that cause the negative reactions we experience with certain substances, giving us allergic reactions. She confessed to having no idea how the suppliers managed to transfer the relevant signature frequencies of the different substances to the vials. This probably reflected in an exorbitant purchase price, I thought to myself.

Once the substance causing the problem had been identified, she would use a special instrument to deliver a mild electromagnetic stimulation to particular acupuncture points. According to her pamphlet, "This improves the body's ability to adapt to the material and achieve homeostasis [stable internal conditions]." If I wanted, I could undergo the procedure myself—there was a special show price of $20. By that point, I had spent the entire morning visiting practitioners of alternative medicine, and my wife had other plans for the afternoon. I therefore arranged to return the following day.

The procedure being offered the next day was for gluten intolerance, and she began by selecting a group of relevant vials, namely gluten, whole wheat flour, white flour, North American hybrid gluten, GMO gluten, GMO wheat, protease (an enzyme that digests proteins), and lysozyme (an enzyme, effective in destroying bacteria, found in several body fluids, including saliva and tears). Setting these to one side for the moment, she said she was going to conduct the muscle-testing procedure.

After sitting down in the chair beside her, I was told to hold up my right hand and make a ring with my thumb and fourth finger, the one next to the little finger. Warning me that she was going to pull them apart, I pressed the tips of my thumb and finger firmly together. Before commencing, she gave me one of the vials to hold against the hand being tested, at about wrist level. She then proceeded to separate the ring while I resisted. After the inevitable breaking of the ring, the test was repeated for the next vial.

When the ring gave way more readily for one particular vile than another, she would comment that this showed I was having a more negative reaction to that particular substance. All that was really happening was that I was not always trying my hardest to resist her pull. The thought crossed my mind to play a game and sometimes let go as if taken by the force, but I resisted the temptation. Having completed the muscle test, it was time for the electromagnetic treatment.

The instrument, about the size of a laptop but much deeper, had several knobs at the front, a couple of electrical cables, and a metallic platform upon which she placed the vials used in the muscle-testing. Before getting started, she asked me to remove my watch, as this could interfere with the internal flow of electricity inside my body. I also had to remove my shoes and socks. Then, passing me one of the cables, she told me to hold the electrode in my right hand, while touching the top of my head with my left hand, palm-side down. She explained how this was connecting positive energy with positive energy. The positive energy came from the outside, through my head, while the negative energy entered through the soles of my feet, or it might have been the other way around. After a few moments, I had to change the position of my left hand so that the palm faced upward. This was now connecting negative to positive energy, or vice versa; it was all rather confusing. While I continued holding the electrode, she took a second electrical cable, which was connected to a small blue light, and held this against one of the acupuncture points on the side of my nose. After about twenty seconds, she changed over to a similar point on the other side of my nose. Next, she applied the light to an acupuncture point in the crook of one of my arms, changing over to a similar point on the other arm several seconds later. The last two acupuncture points to be probed were on the underside of each big toe.

The apparent significance of this procedure was twofold. First, placing the vials of the substances causing the problems onto the metal platform transmitted their signature frequencies to the apparatus. Second, stimulating the acupuncture points with these signature frequencies improved the body's ability to adapt to them, thereby restoring the body's internal balance.

Having completed the electromagnetic treatment, she repeated the muscle-testing procedure. For most of the vials tested, she reported a marked improvement, but two vials were less satisfactory and she wanted to boost my body's response to them. To do this, she placed the two vials on top of the electrical device and repeated the stimulation of the acupuncture points. On retrying the muscle test this time, she found significant improvements—she had restored my internal balance for them.

The entire testing procedure took about half an hour, during which time the congenial lady was amenable to all my questions.

Before entering the booth, I knew I had to have a convincing story to justify my curiosity in her particular kind of alternative medicine. I could not appear too well informed and certainly did not want to appear cynical. Accordingly, I presented myself as someone with an open mind about alternative medicine who wanted to see if it could help with the usual problems of growing older. I also said how my daughter had always been interested in naturopathy and was thinking of pursuing this professionally. My duplicity was not without remorse for such a nice person, but there was no alternative.

When I asked how she got started in bioenergetics, she explained how her young daughter had been suffering from a digestive disorder. Having studied holistic nutrition, she took her to a clinic of alternative medicine, where she was treated by a bioenergetics practitioner. She was so impressed by the improvement in her daughter's condition that she decided to study bioenergetics herself. The program in which she enrolled cost her $10,000, which strikes me as remarkably expensive for a six-day course.

I felt complete empathy for this lady, who, without any understanding of science, had been completely taken in by the entire bioenergetics gibberish. And she is not alone. According to the website of the Institute of Natural Health Technologies, where she studied, there are ten US states and six Canadian provinces with practitioners of BIE. The institute's pseudoscientific ramblings about directing energy to acupuncture points and carrying signature frequencies of problematic substances to achieve homeostasis do not merit serious scientific refutation. Besides, I think they sum it up adequately themselves:

> In order to fully comprehend the new aspects of homeostatic imbalance, one must lay aside the rigid, unequivocal teachings of conventional science.[6]

Readers wishing to read more about this institute of make-believe science may be interested in a pair of articles written by Dianne Sousa for the website *Skeptic North*. With her background in law, she was interested in the legality of the false claim that allergies could be eliminated by bioenergetics. She was also concerned for the potential harm that might result if people with allergic reactions that could lead to anaphylactic shock sought cures for their condition in bioenergetics rather than in conventional medicine. To this end, she lodged a complaint with Health Canada, the federal regulatory body. One of the results of this action was that the Institute's allergists changed their titles from "Registered Holistic Allergists" to "Registered BioEnergetic Practitioners."[7]

Both practitioners and recipients of alternative medicine report health benefits of their treatments, some more convincingly than others. At one of the chiropractic booths I visited, a lady told me about a gentleman who walked into their clinic in such bad shape that he had to use canes. However, after a course of treatment his condition improved so much that he is now playing golf. I had no reason to doubt the veracity of her account, any more than I did that of the bioenergetics lady regarding her daughter. Does this mean that these naturopathic remedies really *do* work? Yes and no. Yes, they work, at least for some recipients, but no, they do not work the way the practitioners say they do. The perceived benefits are attributable to the *placebo effect*, not to the pseudoscience upon which the treatments are based.

Placebos, the proverbial snake oil of traveling medicine shows, have been in use for over two centuries. Elisha Perkins, a Connecticut doctor, patented a device in 1796 that he called *tractors*. His tractors were nothing more than a pair of three-inch-long metal rods tapered to a point at one end. When drawn over affected parts of the body, these thin rods were claimed to bring relief through the electromagnetic influence of the metal. Tractors were used for treating everything from epileptic seizures and gout to pains in all parts of the body, including those caused by scalds and burns. Even George Washington, president of the United States, purchased a set for personal use. Stamped with the name *Perkins Patent Tractors*, they were also sold in England, where they commanded the remarkably high price of five guineas, equivalent today to many hundreds of dollars.[8]

In spite of, or perhaps because of the remarkable success of Perkins and his device, the Connecticut Medical Society revoked his membership, condemning the tractors as "delusive quackery." Others suspected it was trickery, too, including an English physician named John Haygarth. He decided to put this to the test. The tractors supposedly worked because of their special metallic properties, so he made an imitation set from wood that looked identical to the originals. He used the fake tractors on five patients suffering from some malady, leading them to believe they were being treated with the genuine article. Four of them reported feeling better for the treatment. Repeating the experiment the next day with the real tractors, he obtained identical results. He published his findings in 1800 in a work titled *Of the Imagination, as a Cause and a Cure of Disorders of the Body; as Exemplified by Fictitious Tractors, and Epidemical Convulsions.* In a remarkably prophetic statement he wrote,

> An important lesson in physic is here to be learnt, the wonderful and powerful influence of the passions of the mind upon the state and disorder of the body. This is too often overlooked in the cure of diseases.[9]

A great deal has been learned about placebos since Haygarth's time, much of this during the last few decades. Although placebos do not cure disease, they can certainly make people feel better. This is especially true in the treatment of pain. Patients prescribed sugar pills, believing them to be painkillers, frequently report a lessening of their pain and sometimes its complete loss. Similar benefits from the placebo effect have been seen in patients suffering from depression and from Parkinson's disease.

Parkinson's disease is associated with lowered levels of dopamine, a neurotransmitter chemical in the brain that controls movement and coordination. In one study, researchers using PET scans (positron emission tomography) imaged those parts of the brain where dopamine is produced, enabling them to see changes in production levels. One of the drugs used in treating the disease raises dopamine levels, alleviating the symptoms. When volunteer patients were given fake pills, their PET scans showed increased dopamine levels as if they had received the real drug, and their symptoms improved. This clearly shows that the placebo effect is real and not imagined. Similar findings have been documented elsewhere, as when fake pills, thought to lower blood pressure, reduce blood pressure readings.[10]

The reaction to placebos, whether these are pills or some alleged medical procedures, varies among individuals. Some people report great improvements, while others perceive no benefits at all. Such variable responses have been known since Haygarth's time, and recent research indicates that there is a genetic component to account for this. Although these studies are in their infancy, it appears that there is a gene for the response to placebos.[11]

The reaction to a placebo is also determined by the practitioner, and by the physical setting where the procedures take place. People receiving treatments in rooms provided with medical equipment respond more positively than those in unequipped rooms. Similarly, if they are attended to by practitioners who give them lots of attention and consideration, they report greater benefits than those treated by indifferent caregivers.[12]

Placebos are usually associated with positive responses, but they can also bring about negative reactions. This was shown in a study headed by a Harvard professor who was a former acupuncturist. The 270 volunteers in the investigation, all suffering from various pains of the wrist, arm, and shoulder, were divided into two groups, one to receive pain-killing pills and the other acupuncture. They were told that some would be getting real treatments while others would receive sham ones. The sham acupuncture was performed with a device made to look like a real acupuncture needle. When inserted into the volunteers' skin, they could apparently see and feel the needle penetrate. However, the needle was blunt and retracted into the shaft of the handle rather than into their skin as they thought it had.

From the outset, the volunteers were warned of several common side effects of their respective treatments, like temporary pain with the acupuncture and dizziness with the pills. After two weeks, a quarter of the volunteers receiving sham acupuncture and almost a third of those receiving placebo pills reported one or more of the side effects they had been warned about. Both groups experienced similar reductions in their pains but, over the entire eight-week trial period, those receiving sham acupuncture showed significantly greater benefits than those receiving fake pills.[13]

Placebos clearly work, and most doctors have probably prescribed them at some time or another. This raises the issue of ethics in the medical profession, and whether doctors should jeopardize the doctor-patient trust by prescribing fake remedies. It also raises issues in the field of alternative medicine.

I go running most mornings, and my route takes me past a private clinic offering spinal decompression. Here, patients are strapped into an expensive device that exerts large stretching forces on their vertebral column. The pseudoscientific rationale for this is that reducing the pressure on the intervertebral discs draws in water, rehydrating the discs to restore their correct thickness. There is no substantive evidence to support this claim, or to show any clinical benefits of the treatment; the procedure is also expensive.[14] I occasionally catch a glimpse of an early-morning client as I run past the shopfront. The "don't waste your money" thought flits through my mind, but the placebo effect will bring some relief from their backache, so why should I be concerned?

When summer gives way to winter and all those cold and flu commercials appear on TV, along with the usual "clinically proven" patter, I know that many purchasers will experience some benefits because of the placebo effect. I also recognize that the fanciful equipment and surreal explanations given at clinics of alternative medicine will enhance the placebo effect too, benefiting clients and practitioners alike. So where is the harm? The first and obvious source of harm is that by seeking alternate therapies patients are missing the opportunity of alleviating their condition with conventional medicine. In some situations this can prove fatal. Just recently, a case was reported of parents in Alberta who had attempted to treat their lethargic toddler with naturopathic remedies. The child subsequently died of meningitis.[15] There are some other issues too. One concern I have is that well-meaning but scientifically naïve people, like the lady in the bioenergetics booth, can be taken advantage of by potential charlatans. My other concern is that these pseudoscientific treatments help perpetrate scientific ignorance.

One of my amusements, when I have time, is listening to the pitches of door-to-door salesmen. Not too long ago, someone selling filtration systems called. After his opening gambit on the perils of things found in tap water, he asked if I would like him to test my water supply. Silly question.

Setting down his box of scientific equipment on the kitchen counter, he removed a glass flask and filled it with water from the tap. Then, producing a small bottle of liquid from the box, he unscrewed the cap. With a magician's flourish, he added a few drops to the flask and the water turned yellow. "Solutes," he announced grimly. Looking suitably shocked, I stood by while he took out his special measuring device: a probe, connected by wires to an instrument that looked like a voltmeter. This, he explained, would show him the level of solutes in the water.

"See that!" he said, tapping the dial. "There are *lots* of solutes in *this* water."

"Oh no, not *solutes*!" I exclaimed, uttering the dreaded word. "That means I'll have to throw away all the beer and wine in the cellar, along with the fruit juice and milk in the fridge—they're all *full* of solutes."

He did not seem to know that solutes were merely substances dissolved in solvents to produce solutions; like salt, dissolved in water to form brine. But he did get the idea that I was not interested in buying his pointless filter system.

With the disclaimer that I have no problems with people holding religious beliefs, some of my most interesting exchanges over scientific ignorance, and certainly the most entertaining, have involved fundamentalists. The first of these encounters was with a Jehovah's Witness, back in the sixties, soon after I was married. The young man standing on the doorstep was not much older than me, and I invited him inside. We chatted for longer than either of us had expected, and I invited him to come back another day.

One of our long discussions centered upon his belief that gravity was a divine force. I hoped to change his mind with my explanation of a simple experiment, carried out in 1895 by a man named Boys. Charles Vernon Boys's experiment was a repeat, on a much smaller scale, of an experiment carried out almost a century earlier by Henry Cavendish, one of the great experimentalists of his time. Boys's apparatus was essentially a miniature dumbbell, only an inch long, suspended by a fine filament inside an enclosure to eliminate air movements. Bringing two large lead balls close to the two small balls of the dumbbell caused the latter to rotate, due to the attractive forces between opposing doublets of large and small balls. The movement was miniscule, but Boys magnified this by reflecting a narrow beam of light onto a small mirror attached to the dumbbell.[16]

This experiment, which has been repeated many times since, clearly demonstrates that a force of attraction exists between two bodies, due to their masses. My new acquaintance listened intently to my explanation, causing him to change his mind about the divinity of gravity. And that was the last time he called around for coffee and chats. I suspect the elders warned him to stay away from my ungodly influence.

During my battles with creationists, back in the 1980s, I got to know one of them quite well. On one occasion, we were driving back from rural Ontario in a snowstorm, having attended a day-long meeting at a high school on the creation/evolution issue. The 150-mile journey should have taken about three hours, but it was taking considerably longer in the whiteout conditions we were facing, giving us lots of time for discussion.

We got to talking about Noah's Ark and the flood. "He would have to have dropped all the different animals off at their appropriate places after the flood wouldn't he?" I asked. "Over to Africa with the lions and elephants, across to India with the tigers, down to Australia with the kangaroos—like dropping kids off after a hockey match."

"Oh no," said he. "Continental drift would have taken care of all that."

Now, if creationists are going to accept continental drift, they also have to accept radiometric dating. This is because the radiometric dating of rocks provided important information that helped establish the theory of continental drift.[17]

"Hang on," said I, about to point out another problem for him. "If you believe the Earth is only about ten thousand years old, there has not been enough time for the continents to drift apart—they are only moving at a couple of inches a year."

"Ah," said my friend with some conviction, "the rate of drift has slowed down since the time of the flood."

The meeting we had gone to that day had been attended by Dr. Duane Gish, vice president of the Institute for Creation Research, in Texas. He was leading the charge to have the creation science version of how life began, and how it changed over time, taught in schools alongside evolution. A formidable opponent in debates, Gish traveled North American battling evolutionists. He gave his presentation in the morning, immediately before lunch. I had never encountered the man before and sat patiently through his talk. It was easy to see how his slick style and sleight of hand with the facts could win over audiences unfamiliar with science. I had many points to raise concerning his misinterpretations of the fossil record but, since I was the first speaker after lunch, I decided to save my salvo until then.

As we began filing out of the room at the conclusion of the morning session, one of Gish's handlers came sidling across. He wanted to inform me that the big man would not be attending my talk. "Dr. Gish needs to rest before his media appearances this afternoon," he explained. I was not impressed, but my time would come.

Some months later, I received a telephone call. Dr. Gish, the caller explained, would like to meet with me in Toronto for a debate.

By chance, I happened to be watching an episode of a new TV show a few weeks earlier. In this program a contestant would propose a motion, usually something outrageous, and the audience would show their support with a show of hands. The other contestant would then rebut the motion, and the debate would begin. After the allotted time, there would be a second showing of hands. I have long since forgotten what was being debated, but no rational person would have supported the motion, and precious few hands were raised at the start. However, when the debate concluded, the positions had reversed. I realized from this that all creationists had to do to win a debate was fill the audience with supporters who would respond correctly at the appropriate times.

I informed the caller that I would be happy to meet with Dr. Gish, but not for a debate. Instead, we would each give a talk, and these would be followed by a question and answer session. The order of presentations would be determined by the toss of a coin. Leaving him in no doubt that these were the only conditions I would accept, he agreed. Some days later, I received a second telephone call, this one with a warning. The well-intentioned caller, having somehow learned of my forthcoming encounter with Gish, asked if I knew what I was letting myself in for. I assured him that I did.

The confrontation was set to take place in Convocation Hall, at the University of Toronto. Walking across campus on the appointed evening with some of my museum colleagues, we were surprised at what we saw when Convocation Hall came into sight. Everywhere we looked there were parked school buses, all bearing signs of their ecclesiastical affiliations. The creationists were out in full force.

Convocation Hall has a seating capacity close to two thousand and, as I stepped inside, I could see that many of the seats were already occupied. Still carrying the briefcase with my slides and notes, I found one of the organizers and announced my arrival. Taking one of the empty seats in the front row to which he pointed, I sat down and waited.

Those were the days of 35-mm color slides, which were loaded into a carousel that sat atop the projector. I used to use lots of slides in my talks, often exceeding the carousel's capacity of eighty. That night may have been a two-carousel occasion. Speakers would hand over their carousels prior to their presentations, so the projectionist could screen the first few images to ensure everything was working correctly. Because of this, I often placed some alternative slides into the first few slots—usually scenery—to avoid revealing anything to my audience. On this particular occasion, I had substituted some family holiday photos for the usual landscapes.

Shortly after taking my seat, a tall suit glided my way. As I stood to shake his hand, he introduced himself, in his courteous southern way. "You will be speaking first," he announced assuredly.

"No," I corrected, "we'll be tossing a coin as was agreed."

He countered by saying that Dr. Gish had wanted a debate but I had refused, so it was only fair that I should go first.

"I guess we're not going to have any talks tonight then," said I, sitting down again. The suit hurried away.

Returning shortly, accompanied by Gish, the suit produced a coin and the toss was made. I lost. A satisfied Gish went back from whence he had come, leaving me with the handler. The projectionist appeared, and I handed over my slides. I was still standing beside the suit when the first holiday slide flashed onto the screen, followed by the second. "Oh no," I groaned, clapping a hand to my head. "I've brought the wrong slides!"

"What are you going to do?" he asked solemnly, struggling to contain his joy. "You can't go back for the others, there's not enough time."

I shrugged despondently. "Just leave the slides where they are . . . I'll see what I can do with them."

I was pleased with the way things went that night. Jay Ingram, who hosted a national radio program on science at that particular time, said he had never seen Gish under such attack before: "You went right after him!"

Gish was as glib in misconstruing the facts as he had been at the school. But there was something about him that night that made me wonder. He struck me as an intelligent man, and he did have a PhD in biochemistry from Berkeley. Did he *really* believe all the things he was saying, like the Earth being no more than 10,000 years old, not the billions of years as established by radiometric dating? One thing I learned from my interactions with creationists was that no evidence, however compelling, would cause them to accept irrefutable facts over fundamentalist beliefs. Regardless, by evening's end I had my doubts.

From the reaction of the audience it was obvious that the creationists outnumbered the evolutionists by a considerable margin. But there was one person in the audience whose reactions had me completely baffled: Andy Forester. Andy and I had been undergraduates together and shared similar views on creationism. Larger than life, with a stature and voice to match, he was not backward in coming forward. And there he was, sitting in the front row, clapping and cheering and stamping his feet whenever

Gish appeared to make a good point, while booing and shaking his head when I did the same. "What's the Forester up to now?" I thought to myself.

When the last questions had been answered and the evening came to a close, Gish was mobbed by a ring of loyal followers. Meanwhile, I quietly chatted with a few people of similar mind to my own. And that was when I caught sight of Andy, heading straight for Gish. Barreling through the throng, he reached his target. Then, draping an arm around the shoulders of the goblinesque Gish, he seized one of his hands and began shaking it as if he had just found a long-lost friend. I was too far away to hear what he was saying but Gish would have had no trouble with Forester's nose just inches from his face. Moments later and the crowd flew apart, as if a grenade had been released. While official-looking people were shaking their heads, denying any knowledge of who this person was, Andy walked away, smiling.

"What on earth did you say to him?" I asked Forester.

Still smiling, he repeated verbatim what he had said, "Dr. Gish! Dr. Gish! I can't possibly leave here this evening without telling you that you are absolutely . . . full of shit!"

The late Dr. Andrew Forester at his awful best.

CHAPTER 4
PREPARED FOR THE WORST

A long time ago, I discovered that the loneliest place on earth was in front of an audience when one is not properly prepared. I am therefore always ready and well-rehearsed for my talks. However, I don't think I have ever been so well prepared as I was for my presentation to the 2010 annual conference of the Science Teachers' Association of Ontario.

I spent the first sixteen days of the month I'd set aside for preparation in deciding what I was going to cover during the one-hour talk. Flight was the main theme, a subject dealt with in grade six, but I also included topics from later parts of the curriculum, to interest the high-school teachers in the audience. Most of the planning was spent improvising experiments and making models, like a small pair of paper airfoils to show that slender wings generate more lift than broad ones of similar area. The rest of the month was spent rehearsing the presentation. This included demonstrating the hands-on experiments and showing the specimens that I had prepared in advance, like a sample of artificial sedimentary rock that looked and felt like the real thing.

The talk, which was illustrated with sixty slides, was a challenge to complete in the allotted time, and my first rehearsal overran by six minutes. My intention was to keep it well below the hour, to allow for questions; after twenty-three rehearsals I had my timing down to fifty minutes.

As I had so much material to set up in the lecture room before my talk, I arranged to do this at the end of the previous day's presentations. Liz, my wonderfully supportive wife, accompanied me, and we booked into a room at the convention hotel. We retired fairly early that night, but, with the thought of what lay ahead, I did not enjoy the best night's sleep.

Hovering around outside the lecture room next morning as people began filing inside for my talk, I was concerned to see so many young faces in the audience. Would I be booed off the stage when I started ranting about the curriculum? Did I need all this hassle in my retirement years? I had not gone looking for this trouble: it had come and found me. Please let it be over, and soon.

I began by saying how I had once taught high-school science in England, during the sixties, and recounted how my last school was part of the Nuffield Science Teaching Project. In this new approach to teaching science, students spent most of their class time conducting experiments themselves. While skeptical at first, I said how I came to realize that this was the best way for youngsters to learn about science. I used the same hands-on teaching method in my course on vertebrate mechanics that I devised for zoology undergraduates at the University of Toronto.

Turning back to school science, I said that the segment on flight in grade six of the curriculum was an obvious place for stringing together a series of experiments that the eleven-year-olds could conduct themselves.

I started off by holding up three examples of things that flew: a model airplane, a Frisbee, and a bird wing. Knowing that most teachers would likely not have a bird wing at their disposal, I explained how dead birds can often be found outdoors, and how easy it was to remove a wing and prepare it for examination by their students.[1]

Picking up each specimen in turn, I pointed out that they all had one feature in common: a convexly curved upper surface. After using a sheet of paper to demonstrate how lift is generated when air flows over a convexly curved surface, I demonstrated the toilet-roll-tube airfoil. Simple airfoils like this—convex above and concave below—are found in the wings of birds, bats, and insects. Similar wings were used in the earliest attempts at heavier-than-air flight, inspired by human observations of birds and their wings. One of the pioneers of such flights was Otto Lilienthal (1848–1896). As an image of a man harnessed to a large pair of bird-like wings appeared on the screen, I described how he had made hundreds of successful gliding flights by launching himself off hills in his native Germany. His longest flights were in the order of 250 meters (almost 300 yards), a record unsurpassed at the time of his premature death in a fatal crash. The dying words of this remarkable man were, “sacrifices must be made.”[2]

Figure 4.1: Photograph of Otto Lilienthal, about to launch himself from a hilltop.

Figure 4.2: Photograph of the *Wright Flyer 1*.

The Wright brothers' aircraft, the *Wright Flyer 1*, had similarly open-structured wings, as did the earliest biplanes and monoplanes. Later aircraft of the early 1900s had enclosed wings with flat undersurfaces.

Figure 4.3: A US-built DH-4 (originally built by the UK company De Haviland). The upper and lower surfaces of the thin wings follow similar contours.

Figure 4.4: A Boeing Model 40 A. The wings are enclosed by a flat underside, giving them greater depth for improved lift.

I demonstrated a tethered flying model of such a wing, made from a narrow rectangle of paper. Construction is a simple matter of folding the paper in half lengthwise, and then reducing the width of one of the halves by bending over a narrow flap a few millimeters wide. This half of the rectangle will become the lower surface of the wing, and the narrow fold-down flap serves for tucking in the free edge of the other half, which will become the upper surface. As the upper surface is wider than the lower one, it bulges up, forming the convexly curved profile. Four cotton threads of equal length are taped at the four corners of the airfoil, just inside the creases, to act as tethers. The model is completed by securing the narrow flap with small pieces of Scotch tape. I demonstrated the superiority of the paper airfoils over the toilet-roll variety by using a small keyboard blower to fly one, instead of the more powerful hairdryer.

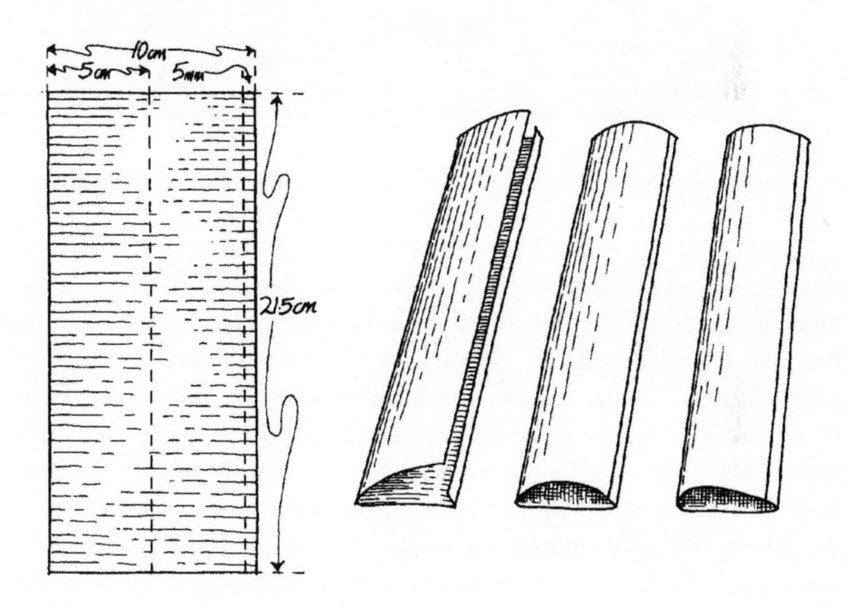

Figure 4.5: Stages in making an airfoil from a strip of paper.

Figure 4.6: The completed paper airfoil is tethered to a piece of cardboard.

The lift generated by an airfoil increases with camber, which may be thought of as a combination of its curvature and thickness.[3] I demonstrated this with a pair of tethered airfoils made from a toilet-roll tube that had been cut in half lengthwise. One half was left, unmodified, as the high-cambered airfoil. The other had its camber significantly reduced by placing it, convex-side up, on a flat surface and repeatedly pressing it down by hand. This reduced its height from ¾ inch (18 mm) to ½ inch (12 mm). Each airfoil was mounted on a sheet of cardboard, with enough slack in the tethers to allow them to rise a couple of inches into the air. Each airfoil was tested, in turn, with a hairdryer held about six inches (15 cm) in front of it. The dryer was set at low speed, and the nozzle was aimed slightly downward. The high-cambered airfoil immediately took off, while the other failed to lift completely clear of the base.

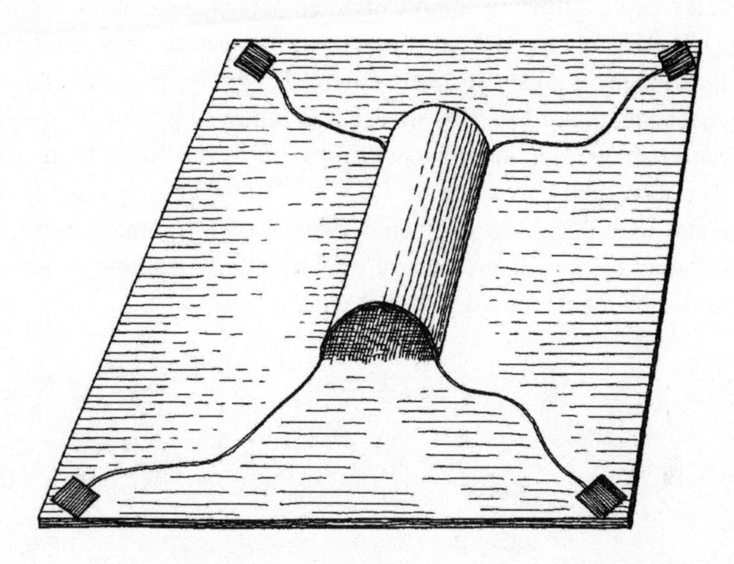

Figure 4.7: The unmodified half of the cardboard tube, showing its high camber.

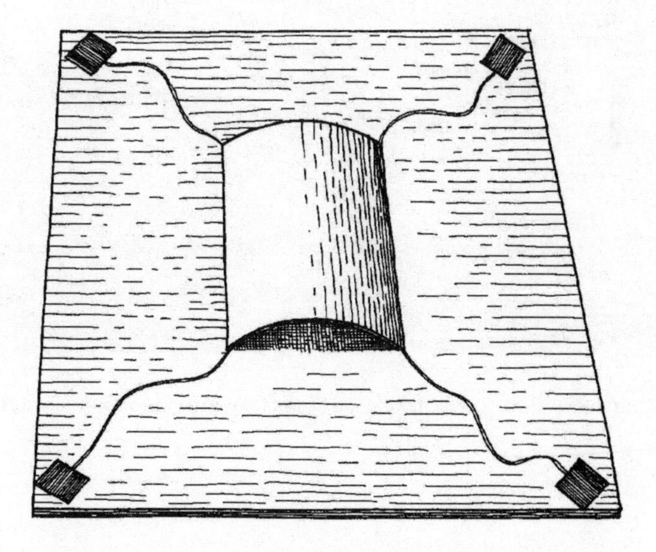

Figure 4.8: The flattened half of the cardboard tube, showing its reduced camber.

While explaining how high-cambered wings were used for heavy lifting, a Boeing 747 flashed onto the screen. The leading-edge and trailing-edge flaps of the wings were fully extended, giving the wing maximum camber for maximum lift. This image was followed by one of an osprey (fish eagle) showing its deep broad wings, necessary for carrying off its prey. The next image was that of a CT-114 Tutor aircraft, a lightly built twin-seater with fairly low cambered wing, used for training pilots. This was followed by a photograph of a hummingbird, captured in hovering flight and showing its remarkably thin wings.

Figure 4.9: A Boeing 747 taking off, flaps fully extended for maximum lift.

Figure 4.10: An osprey, showing its high-cambered wings.

Figure 4.11: The CT-114 Tutor jet is lightly built and has only modestly cambered wings.

Figure 4.12: A hummingbird, showing its remarkably thin wings.

The lift generated when a wing sweeps through the air is always accompanied by drag, a force that opposes forward motion, thereby wasting energy. I demonstrated this by holding a sheet of foam core vertically in my hand and forcefully moving it from one side to the other. As I commented, those sitting closest to me could feel the draft due to the air being displaced and thrown into turbulence. The drag generated by the board was obvious from the resistance I could feel as I pushed it forward. Repeating the demonstration with the board held horizontally caused it to slice through the air, seemingly without any resistance; nobody in the audience could feel any draft. The air was now flowing smoothly over the board, and I explained that such flow was described as being *laminar*.

I commented that most of us have seen laminar and turbulent flow when using a tap. If a tap is turned on gently, the water flows in a smooth column—the flow is laminar. Turning the tap on full changes the glassy flow into a rough and tumbling torrent; the flow is now turbulent. Reducing turbulence reduces drag, and anything that can be done to promote laminar flow in bodies moving through air or water reduces the energy required to propel them. For ships and planes this translates into lower fuel costs; for birds and fishes the savings are in food consumption. Enhancing laminar flow also enables higher speeds to be reached.

The optimum outline for minimizing drag is the streamline, a teardrop-shaped structure that is roundly pointed at the front and gently tapered toward the back. This shape is usually associated with fast cars and planes.

At this point in the presentation, a diagram appeared on the screen depicting the flow of fluid around a disc, a sphere, and a streamline, all of

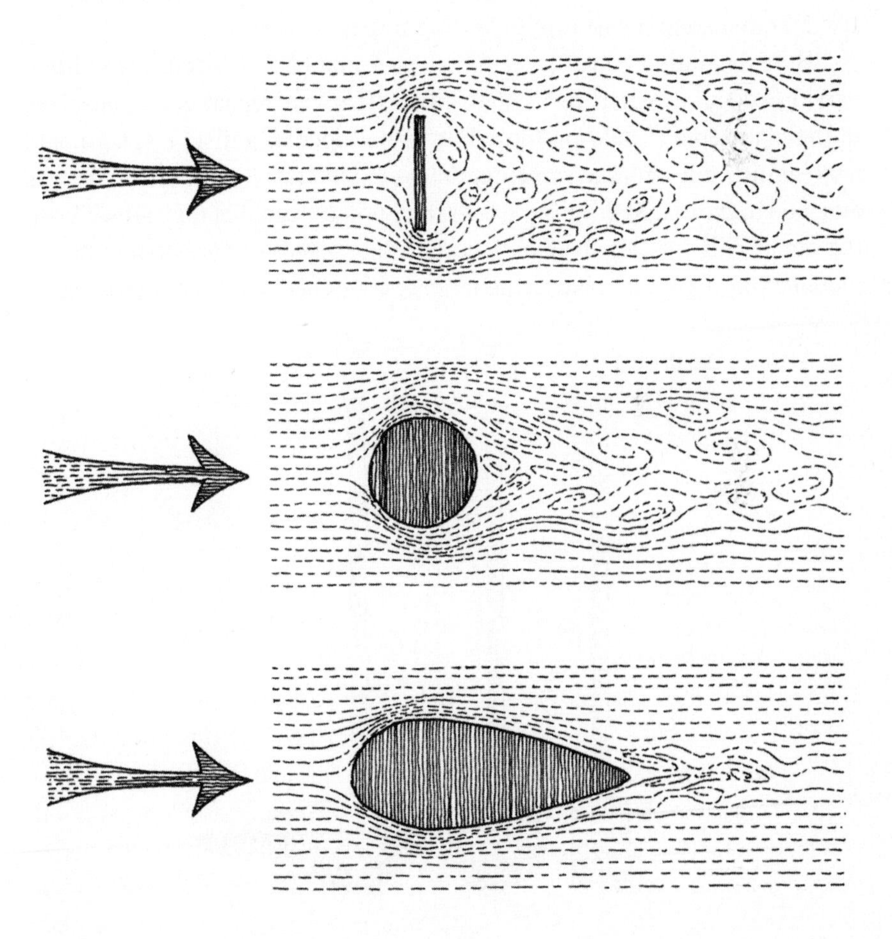

Figure 4.13: Diagram showing how the flow of fluid around a disc (top) is so turbulent that it creates a much wider wake than that of a sphere (middle). The streamline creates the smallest wake.

the same cross-sectional area.[4] I explained that when a fluid separates into streams to pass around a disc, it breaks away at the edges and becomes turbulent. This creates a significant wake that trails behind, generating a large drag force. The flow around a sphere is less turbulent and the fluid does not break away until it has passed beyond the widest point. The wake is therefore narrower and the drag force accordingly less than for a disc. The flow around a streamline is essentially laminar throughout, with no separation from the surface and barely any wake, so the drag is minimal. Such differences in performance can be seen by testing models of the three shapes in a wind tunnel and measuring their drag forces.

By substituting a hairdryer for a wind tunnel, I devised a simplified version of these experiments that youngsters can repeat for themselves in the classroom. I held up three Plasticine models: a disc, a sphere, and a streamline, all with the same diameter (1¼ inches; 3 cm). Each one was attached to the end of drinking straw, and I pointed out that they all weighed the same. This was achieved by incorporating pieces of Styrofoam into the models; the streamline, being the largest, contained the most Styrofoam.

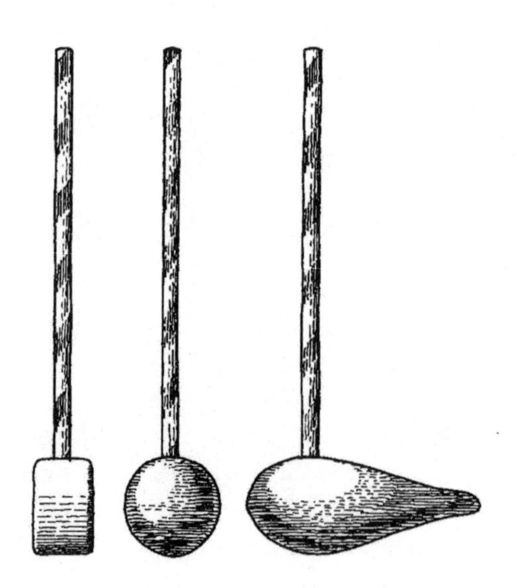

Figure 4.14: The three Plasticine test models, attached to their drinking straw supports.

The next item I demonstrated was an empty 2-liter milk carton, modified to support each of the three models. All I had done was cut a cork in half and glue it to the top of one side of the carton, exactly in the middle. A rectangle of paper with a prominent black line drawn down the center was then glued to the side, positioned so that the vertical lined up with the center of the cork. Last, some sand was added to the carton to prevent it from moving during the experiment.

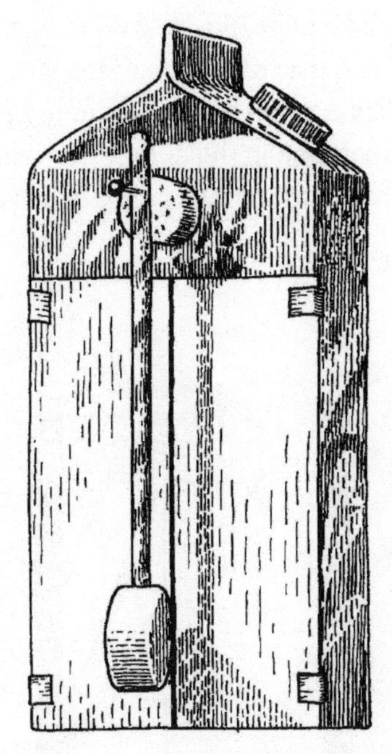

Figure 4.15: The test-bed for the models: a milk carton, partly filled with sand for ballast, with a slice of cork glued to one side for attaching the model to be tested.

Before each model was attached to the carton, its supporting straw was speared with a quilting pin—one of those long pins with the colored heads. The pin needs to be placed about an inch (2 cm) from the end of the straw, so that the model clears the surface when the pin is attached to the cork. When mounting the model on the milk carton, the pin is positioned so that the straw lines up with the black center line. Once the model has been attached, the hairdryer has to be positioned to aim air at its front, directly in line with the direction of swing.

The hairdryer is set up so that the nozzle lies parallel with the top of the table, at the same height above the surface as the model. The nozzle of the hairdryer I was using was much less than the width of the body, but I overcame the problem by attaching a couple of stacks of Post-it notes to the nozzle with an elastic band, to keep it parallel with the table. I used the hairdryer at full power, with the nozzle set a fixed distance, 4½ inches (11 cm), from the center line on the milk carton.

The differences in drag between the models were quantified by measuring the differences in their displacements when being blown by the hairdryer. Using a pair of dividers, this was done by measuring the distance between the center line and the front edge of the straw at the point where it attached to the model. Time was a constraint in my presentation, so I only used the disc and the streamline in the demonstration. The displacement of the streamline was half that of the disc, clearly showing its effectiveness in reducing drag.

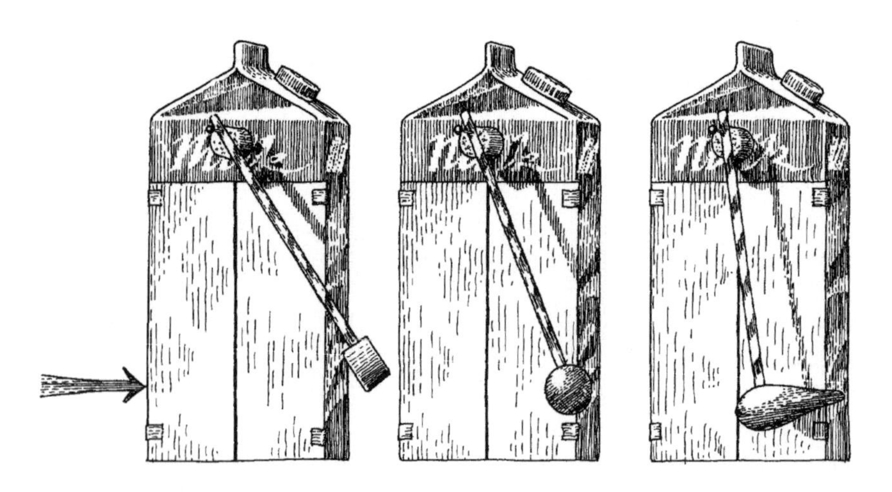

Figure 4.16: The disc (left) creates the largest drag, causing it to be displaced the farthest from the vertical center line on the milk carton.

The drag on an object increases with the density of the fluid. If you held out the flat of your hand while walking at a fast pace you would be unlikely to detect any drag. However, if you presented the same hand over the side of a canoe being paddled at the same speed, the water, which is over 800 times denser than air, would exert a considerable drag. As the image of a bathtub appeared on the screen, I told the teachers how their students could readily repeat the drag experiments at home, by moving their models through water. Swishing the disc across the bath at full speed generates such a large drag force that the straw is kinked backward. Repeating with the streamline generates such minimal drag that the straw remains unkinked, no matter how fast it is moved. Students will also be able to see the smoothness of the flow over the streamline, and the narrowness of the wake, compared to the great turbulence in the water caused by the disc. Another benefit of conducting drag experiments in water is that the models do not have to be of the same weight. Examples of streamlines in the natural and man-made worlds were illustrated with images from the Internet that now appeared on the screen.

Figure 4.17: This superb photograph of an Arctic tern captures the streamline shape of the body to perfection.

Figure 4.18: The streamline shape of a canoe enables it to slip through the water with minimal drag.

Figure 4.19: The bodies of tunas are so beautifully streamlined that their pectoral fins—particularly long in the albacore tuna shown here—fit into shallow recesses at the sides of the body, making the surface perfectly smooth.

Figure 4.20: This World War II Spitfire has a streamlined fuselage, but it is nowhere near as impressive as that of the Arctic tern.

Returning to the subject of flight, I compared the flapping flight of familiar birds seen in backyards, like sparrows and grackles, with the seemingly effortless soaring[5] of larger birds, like gulls and hawks. Small birds have to continuously flap their wings to remain airborne, using muscle power that is energetically expensive. While feasible for small birds, this becomes untenable for large ones. This is because as things get bigger their masses increase exponentially, rising with the cube of the size difference.[6] Doubling the size of an object therefore increases its mass by eight. I illustrated this relationship with some familiar objects. The first image that appeared on the screen was a pair of plastic containers, filled with a popular household cleaner. One was marginally taller than its partner, at 24 cm, compared to 19 cm. The increase in size was only $24/19 = 1.26$, but $(1.26)^3 = 2$. Checking the labels revealed that one contained 500 ml and the other 250 ml. The same relationship can be demonstrated for any pairs of solid objects—from onions and tomatoes to people and pets—provided they are of similar shape.

Large birds have to flap their wings to become airborne, after which they can soar by exploiting air currents, with relatively small expenditures of energy. Sometimes large birds can reduce these takeoff costs by launching themselves from trees, or by exploiting other aspects of their environment. Many years ago, in the Galapagos Islands, I saw a Magnificent Frigatebird take off effortlessly, simply by facing into an onshore wind and spreading its slender two-meter (6 feet) wings. On another occasion, this time in New Zealand, I saw a Royal Albatross attempting to take off by running, seemingly as fast as it could, across the high ground overlooking the sea. The albatross was using a runway that had been cut in the long grass by local conservationists to assist these birds to become airborne. The bird, weighing up to about 22 pounds (10 kg) with a wingspan of around 10 feet (3 m), was flapping furiously, but it had to abort the takeoff before running out of ground. The attempt was repeated a few more times before the bird finally gave up, presumably to wait for more favorable onshore winds.

Ideally, a wing should generate maximum lift for minimum drag. The Wright brothers were well aware of this and built a wind tunnel for testing models of wings. They measured the lift and drag for more than one hundred models, expressing the results as the ratio of lift/drag. This was the first time anyone had ever conducted such an experiment. After discovering that long slender wings yielded higher ratios than short broad ones of similar surface areas, they built the *Wright Flyer* accordingly. The slender-

Figure 4.21: Photograph of a facsimile of the Wright brothers' wind tunnel.

ness of a wing is expressed by the *aspect ratio*, obtained by dividing the length of the wing by its width. The aspect ratio of their powered aircraft was just over six, which was twice that of the first two gliders they built.[7]

To demonstrate the superior lifting capacity of wings with high aspect ratios, I had constructed a pair of small paper models. These wings had the same surface areas and the same maximum depths, but differed in their aspect ratios. One wing was square, the other long and narrow.[8] The construction of the wings had involved much trial and error and, because they were so small—the narrow one was only ¾ inch (18 mm) wide—I had to use the smallest keyboard blower I could find to test them.[9] Holding the small blower a few inches from the narrow wing caused it to take off and fly. Repeating this for the other model caused the leading edge to rise, but the wing failed to lift off from the surface. The best exemplars of high aspect-ratio wings in human aeronautics are gliders, sometimes called sailplanes. In the natural world, the albatross tops the list. These magnificent birds, with their long slender wings, spend most of their lives in the air, traveling thousands of miles across trackless seas in the southern hemisphere, foraging for food. Satellite tracking devices attached to individual birds have recorded them covering distances of up 15,000 km (9,000 miles) during single foraging trips.[10]

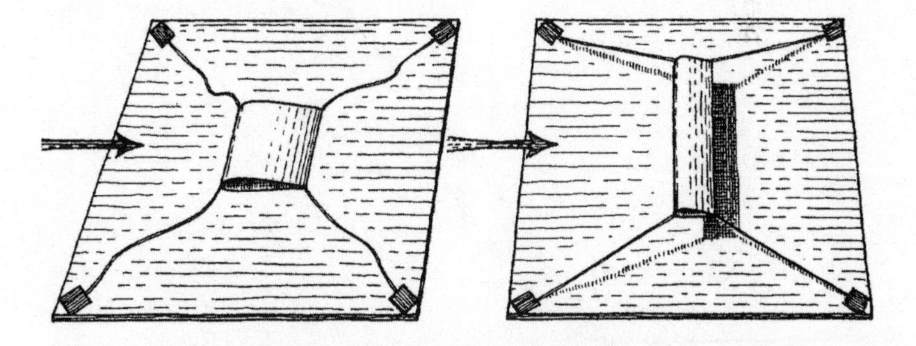

Figure 4.22: These small paper airfoils have the same surface areas and the same depths, but differ widely in shape. Because of the superior lifting capacity of slender airfoils, the one on the right readily took off when gently blown, while the other failed to become airborne.

Figure 4.23: Gliders have remarkably slender wings to maximize the lift received from exploiting moving air flows.

Figure 4.24: This black-browed albatross, like other members of this family, have the highest aspect-ratio wings of all birds.

One of the biggest problems the Wright brothers faced was controlling their aircraft. The *Wright Flyer* was very unstable, and the methods they used for changing directions and for restoring stability were not used in later aircraft. Before directing my audience's attention to how aircraft are controlled, I had to describe the three directions of movement—*pitch*, *roll*, and *yaw*. I did this by holding up a model aircraft and going through the motions. Pitch is the up and down movement, when the nose lifts or drops. Roll is when one wing drops or rises, causing the fuselage to roll about its long axis. Yaw is the side-to-side motion, when the nose swings toward the left or right. For white-knuckle fliers like my oldest daughter, this would probably be the most disturbing movement to experience during a flight.

Picking up the foam core sheet used to demonstrate drag, I held it in one hand and tilted it up so that when I moved it through the air, the leading edge was higher than the trailing edge. This, I explained, was now an inclined plane,[11] and the angle of the tilt is referred to as the *angle of attack*. Using my free hand, I traced the arc between the lower surface of the board and the horizontal. At positive angles of attack like this, I explained, the inclined plane generates an upthrust. When I swished it rapidly through the air, the sheet naturally followed an upward path. Upthrust is generated because the plane deflects air downward, away from the lower surface, resulting in an equal and opposite force acting upward. Tilting the plane the other way so that the leading edge is lower than the trailing edge generates a downthrust; the inclined plane now has a negative angle of attack.

Swishing a board through the air demonstrates the principle of inclined planes but does not allow for any comparisons to be made between the forces being generated. For this, I demonstrated a simple experiment I had devised using inclined planes made from a throw-away aluminium pie plate. Each plane was made by cutting out a strip measuring 3 × 1 inches (7.5 × 2.5 cm), with a small hole made in the center just large enough to push through a drinking straw.[12] Cutting off a 2 inch (5 cm) section of drinking straw, this was pushed halfway through the plane and fixed in place with a ring of Plasticine, attached to the underside of the plane. After setting the plane at an angle of about 30°, it was firmly secured by flaring and smoothing the Plasticine plug against the aluminium.[13]

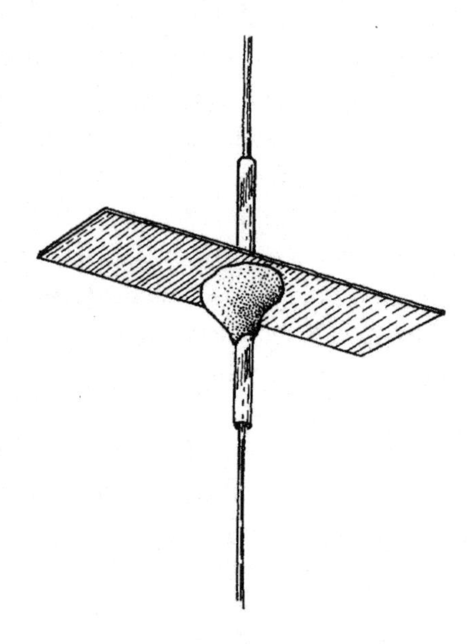

Figure 4.25: This small inclined plane, cut out from a disposable pie plate, is attached to a short length of drinking straw by a plug of Plasticine. The plane is free to ride up and down on a shish-kebab stick, anchored vertically in a slab of Plasticine.

The inclined plane was mounted on a wooden shish-kebab stick, embedded in a slab of Plasticine to keep it vertical. The slab had to be large enough to withstand the outflow from a hairdryer.[14] Two additional sticks had to be added on either side of the inclined plane, almost in contact with the edge, to prevent it from spinning when blown by the hairdryer. Depending on which way the inclined plane faced, it could be set up with a positive or a negative angle of attack. When set with a positive angle, the plane rises when the dryer is turned on. By moving the hairdryer up or down, the plane can be made to hover at any level. When placed at a negative angle of attack, the plane has to be elevated from the base, by about 3 inches (7 cm), and held in place by a thin strand of elastic, attached at the top of the center support. Such thin elastic (less than 1 mm thick) can be teased from a piece of elastic from a needlework store, or from the elastic band of a discarded item of clothing.

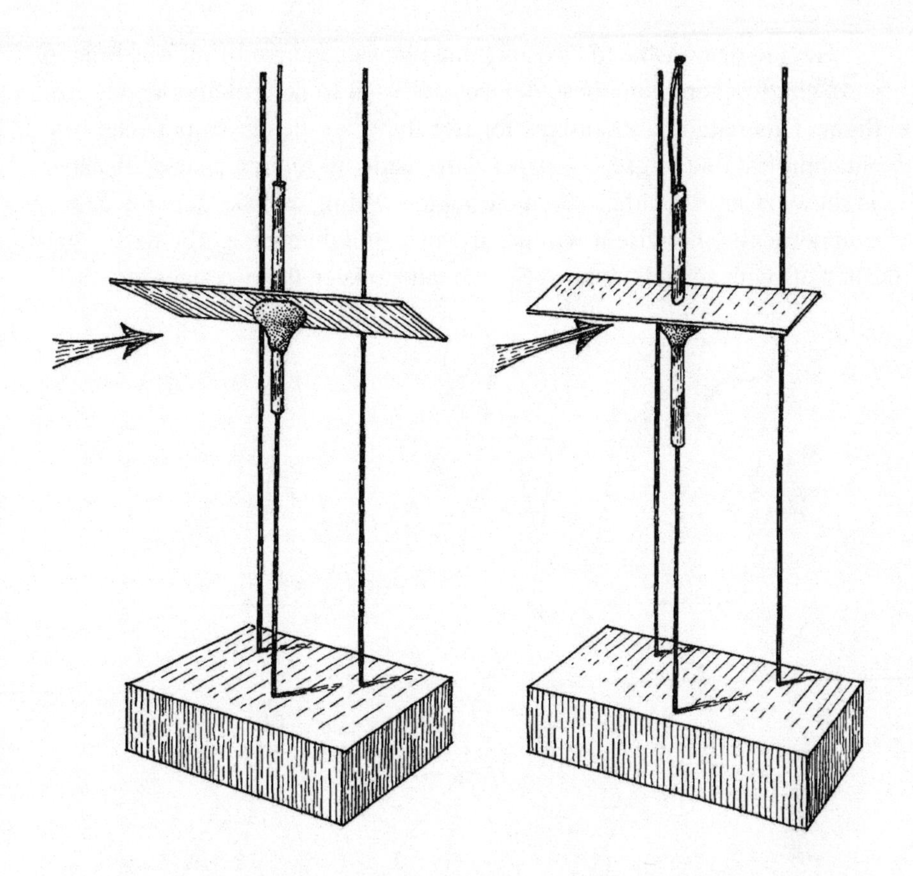

Figure 4.26: Additional shish-kebab sticks are mounted on either side of the inclined plane to prevent it from spinning when blown by a hairdryer. When set with an upward tilt (shown on the left) it rises when blown. The reverse is true for the other inclined plane, so it is supported part-way up, using a thin strand of elastic.

By constructing a series of inclined planes set at different angles of attack, students can explore how lift changes with the degree of inclination. This, of course, requires a way of comparing the lift forces generated by different inclined planes. One way to do this is to see how far away the hairdryer can be placed before the plane no longer takes off.

Having demonstrated how inclined planes generate lift, it was time to show the teachers how these devices are used to control and to stabilize flight. This required an aircraft for test flying in the classroom. The best and simplest one to use is a paper dart, made by folding a sheet of paper as shown. Care should be taken during the folding to make sure the dart is symmetrical; otherwise it will not fly on a straight course. The dart is finished off with a small piece of Scotch tape to keep the two sides together.

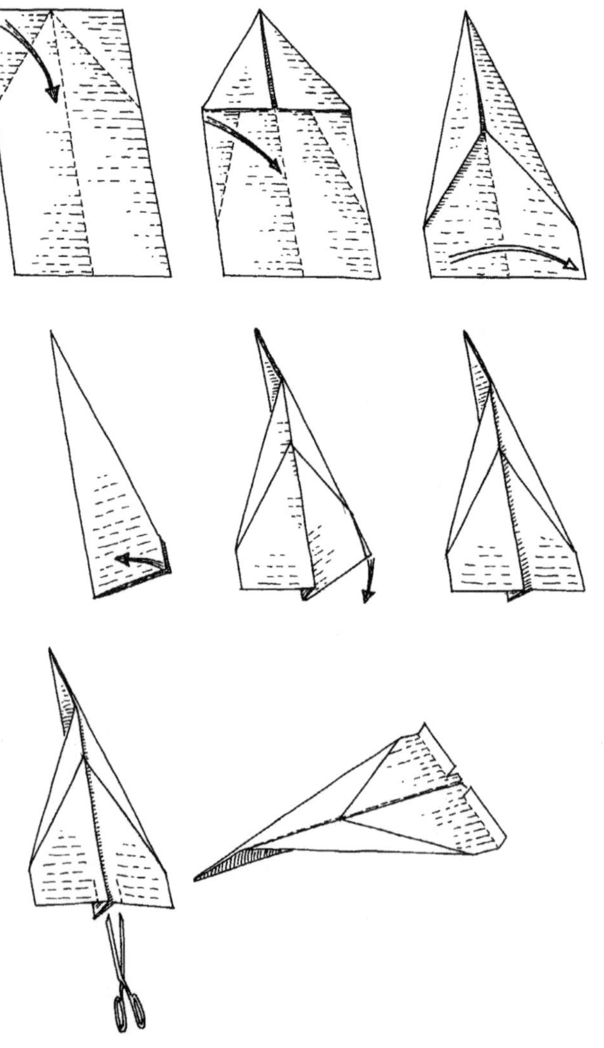

Figure 4.27: The steps involved in making a paper dart.

The inclined planes are formed by cutting a pair of slits, of equal length, at the rear of the dart, on either side of the keel—the central fold that runs along the length of the dart. Tilting the planes up, as show in the finished dart (see figure 4.27, bottom right), sets them at a negative angle of attack. This generates a downthrust when the dart is in flight, causing the nose to pitch up. The reverse is true when the planes are bent down. If one plane is tilted up and the other down, they generate forces in opposite directions, causing the aircraft to roll in flight.

Because of the time constraints of my presentation, I had made three paper darts with their inclined planes already set in these different configurations. Launching each one in turn was a convincing demonstration of how inclined planes could be used to effect changes in direction. And then I showed them a fourth dart, which had been made by one of my grandsons. This dart differed from the others in being without inclined planes on its trailing edge. Instead, the wingtips had been bent upward, forming a pair of vertical fins, set at the rear end of the dart. When I launched this dart across the room, it flew in a remarkably straight line: the aircraft was strikingly stable. I explained how the fins were acting as inclined planes, like the vertical part of an airplane tail (called the vertical stabilizer). Seen from above, it was easy to see how both would generate thrusts if the dart yawed away from a straight path, setting it back in line. My grandson was only seven when he made the dart. As I commented to the teachers, this is an excellent example of what youngsters can achieve if motivated and properly directed.

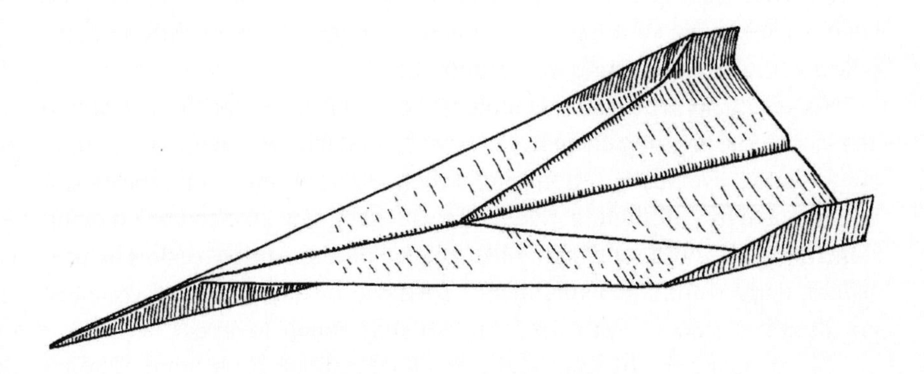

Figure 4.28: A paper dart with its wingtips bent into vertical fins is remarkably stable in flight.

Pausing at this point, I announced that I needed to check the curriculum to see how well I had covered the topic of flight. As images of the relevant sections appeared on the screen, I commented on the absurdity of so much of the content. The things I had omitted ranged from assessing "the benefits and costs of aviation technology . . . [from] . . . the perspectives of farmers, airline workers, doctors, home owners [and] tour operators" to considering how "air travel . . . increases the risk of spreading infectious diseases . . ."[15] From the raucous reaction of the audience, it was apparent that they held similar views to my own. I admitted that if I were teaching a section on flight in the classroom, my students would fail to make the grade because I had left out so much of the curriculum. However, they would have a sound understanding of the science of flight through learning from their own hands-on experience.

Moving on to some topics from grade twelve, I turned my attention to rocks and the fossil record. As photographs of the stratified sea cliffs at Lyme Regis filled the screen, I described how it was along this stretch of early Jurassic rock that Mary Anning, the pioneer English fossilist of the early 1800s, unearthed the first ichthyosaur, at the age of twelve.[16] Ichthyosaurs, fishlike reptiles that lived in the sea, were contemporaneous with dinosaurs.

I explained how sedimentary rocks were formed at the bottom of seas, lakes, and rivers by the accumulation of sediments raining down from above. Picking up an empty jam jar, I said that this could be replicated in the classroom with a simple experiment. All you had to do was pour in some water, add some soil and sand, screw on the lid, and give it a good shake. After leaving it to settle for about half an hour, you were left with bands of sediment, stratified according to the densities of the different particles, with the sand resting on the bottom.

As sediments settled and accumulated over the millennia, the pressure of all the material pressing down from above caused the lower layers to become consolidated, eventually forming solid rock. This consolidation process can be simulated by substituting glue for pressure and time, the end result being something that looks remarkably like a piece of sedimentary rock. The procedure, derived from an experiment I conducted with my second-youngest grandson half a dozen years ago, is satisfyingly simple to repeat.

Equal quantities of sand and soil (1 tablespoon each), and 5 tablespoons of water are thoroughly shaken up inside a closed glass jar. The muddy slurry is then allowed to settle for about five minutes. After skimming off any floating material with a teaspoon, add 2 teaspoons of glue[17] and thoroughly shake again. Quickly pour the contents into a paper or Styrofoam cup, swiftly scraping out any sludge left behind.

After leaving the cup and contents overnight to settle, it is placed on a large plate to catch the water that has to be drained off. Depending on which glue was used, there may be a jellylike layer floating on the surface.[18] Using a teaspoon, this layer must be removed before draining begins, scraping around the edges for the last dregs. Draining begins by stabbing a hole through the side of the cup with a sharp pencil, just below the level of the water. The process is then repeated with progressively lower holes, the last one being made at the surface of the soft deposit. Taking care not to disturb the sediment, the cup is slightly tilted to drain off the last of the water. The cup is then set aside to dry out for two days, or for longer if there are any traces of liquid on the surface. After drying, place the cup into a freezer for three hours to solidify the contents. The last step is to cut off the top of the cup, down to the level of the sediments, and carefully peel or cut away the remainder. The artificial sedimentary rock is left to dry for two days before being handled.

Figure 4.29: Artificial sedimentary rock.

From sedimentary rocks, I turned my attention to the fossils that are sometimes found within them. The photograph of a hadrosaurian dinosaur appeared on the screen, one that had been mounted standing in a fossil trackway. The footprints had not been made by this particular specimen, but they were hadrosaurian and had been collected from the same dinosaur locality, in western Canada. These tracks, imprinted on the ground over sixty-five million years ago by some passing hadrosaur, had not been destroyed and lost along with countless others. By some fortunate happenstance, the sediments had become consolidated, preserving the prints forever.[19]

Figure 4.30: A hadrosaurian dinosaur skeleton, as displayed in the former dinosaur gallery at the Royal Ontario Museum, in Toronto.

Figure 4.31: A close-up view of one of the hind feet of the hadrosaur, showing the imprint of a hadrosaurian foot.

Students can simulate this process in a take-home project to make their own preserved animal footprint. All that is needed is some sand, a shallow container like a sardine can or cut-down paper cup, some clear nail polish, and a cooperative cat. After filling the container with sand, one of the cat's paws is gently pressed into this to make a footprint.[20] Clear nail polish is then dripped into the depression. This rapidly soaks into the sand and dries. After leaving the impression to harden for a couple of hours, it can be removed from the sand and handled. This project can be modified for the classroom by substituting a finger or some inanimate object for a cat's paw. At this point, I showed the teachers a sacred relic: the footprint of the family cat that had died many years earlier.

The next image to appear on the screen was the vertebral region of another hadrosaur skeleton, this one showing a patch of sandstone overlying some of the vertebrae. A close-up of the patch revealed a textured pattern. I explained that this was a natural cast of the dinosaur's skin, the pebbled patterning showing what hadrosaur hide would have looked like in life. This contrasted with the fossil footprints seen before, which were marks left behind by the feet rather than being what the bottoms of the feet would have looked like. I told the teachers to visualize the body of a hadrosaur lying on the ground, imprinting an impression of its pebbly hide into the soil. Then imagine the impression becoming filled in with a thin layer of sand, and the sand, over time, becoming consolidated into sandstone. What remains is a positive replica of the skin, in contrast to a negative impression of the foot.

Figure 4.32: A patch of sandstone overlying part of the vertebral column of a hadrosaurian dinosaur.

Students can replicate this in the classroom using a husked corncob as a substitute for hadrosaur hide. The other items needed are a shallow container, large enough to hold the cob, enough damp soil to fill it, about half a cup of clean dry sand, clear nail polish, an unopened soda can, a plastic knife, and an old toothbrush. After filling the container with soil, it is rolled flat with the can. The corncob is then firmly pressed into the soil, rocking it slightly from side to side to leave a good impression. Once the cob has been removed, some of the sand is sprinkled into the impression to form a thin layer. When satisfied that the entire impression has been covered, the rest of the sand can be added quickly, forming a layer of uniform thickness. Nail polish is then dripped evenly into the sand to consolidate the grains and left to dry until the next day. After checking to make sure the artificial sandstone feels hard, the soil is loosened with the knife and lifted free. The soil is then carefully scraped away, but if it feels soft it is left to harden for another day. The last of the soil is brushed off with the toothbrush. If necessary, cold water can be trickled onto it while gently rubbing with the fingers.

Figure 4.33: A close-up of the sandstone patch reveals that it is a natural cast of the dinosaur's skin.

Fossils are rare, and paleontologists can spend days and weeks prospecting promising rock exposures before finding any. This is because the chances of preservation are exceedingly small, due to the fact that most animals are scavenged soon after death, their bones being scattered, broken, and lost. The best chances of preservation occur when bodies are rapidly buried, thereby protecting the hard parts—the bones and teeth—from destruction. I then described a long-term experiment that can be conducted on school grounds, or in backyards, to see what happens to bones left in the outdoor environment.

Leftover chicken bones are used in this experiment, and those from the leg (which are usually more readily available) are probably best. A set of four bones is required, all of the same element; I used femora. After boiling them for a short while and leaving them to cool, they are scrubbed clean. One of the bones is securely tied to a few feet of strong string, the other end being tied to a wooden stake, which is hammered into the ground. The second bone is placed beside it. The third bone is buried, at least a foot deep, and the soil firmly packed down upon it. The fourth bone is placed somewhere safe indoors, as a control, for comparing with the other bones at the end of the experiment. Ideally, the investigation should be conducted over a full year.

I showed the teachers the end results of my experiment with four chicken femora. The un-tethered bone had disappeared after the first night. The tethered one looked like the control bone, but the surface felt rough rather than smooth. This was due to weathering, likely from the effects of the sun. The buried bone still felt smooth, but the color was a little different, due to the leaching-in of elements from the soil. Similar discoloration occurs in fossils. For example, the bones of hadrosaurs, and other dinosaurs from the Alberta badlands, are brown, due to the iron that diffused into the bone from the surrounding sediments.

This concluded my presentation to the teachers and many of them stayed behind to chat. Some wanted to talk about the experiments, but others wanted to share their concerns about the senseless curriculum. Had there been any teachers from the United States in the audience, they too would likely have shared their frustrations over the directives they were obliged to follow in teaching science.

CHAPTER 5

THE BEST LAID PLANS

Having been raised in the United Kingdom, where there were national standards for science education, I was surprised to find how regionalized the science standards were in the United States. While I expected each state to have its own goals and approaches to teaching science, I did not think this would extend to the various school districts within a state. It seems odd that teachers and students at a given school can have a different science curriculum from those nearby who just happen to be in another district.

This parochialism makes the Next Generation Science Standards (NGSS)—the new set of US standards for the technological age—all the more remarkable because they were developed with the assistance of twenty-six participating states. The NGSS, a joint venture between the National Research Council (NRC), the National Science Teachers Association (NSTA), the American Association for the Advancement of Science (AAAS), and Achieve (an independent educational organization), was to provide a set of expectations and goals for students to achieve during their school careers. This is intended to "stimulate and build interest in STEM" (science, technology, engineering, and mathematics) and "better prepare high school graduates for the rigors of college and careers."[1] This, in turn, would provide employers with a pool of potential employees who could bring valuable science-based skills to the workplace. Incidentally, while STEM subjects are taught as an integrated entity in the United States, they are usually taught separately in the United Kingdom. President Obama, a prominent advocate of STEM education, said, "One of the things that I've been focused on as President is how we create an all-hands-on-deck approach to science, technology, engineering, and math."[2]

Since the time that the NGSS was released, in April 2013, twelve states and the District of Columbia have committed to its acceptance. Formulating the NGSS was a two-step process. The first phase was the development of *A Framework for K-12 Science Education*, described as “a framework that articulates a broad set of expectations for students in science.”[3] The *Framework* formed the basis around which the NGSS developed, so this will be the subject of the present chapter.

The *Framework* document runs for almost four hundred pages, and it would be unrealistic to attempt to cover its entirety here. Given the challenging nature of some of the writing, this would also make for a daunting read. I will accordingly focus on the salient points, those having a profound bearing on how science is to be taught in US classrooms.

Work on the *Framework* began in 2010, which seemed an auspicious time given the large number of states adopting the Common Core State Standards for mathematics and English. Although national documents for science standards were already available from the previous decade, it was considered time for something new: “Not only has science progressed, but the education community has learned important lessons . . . and there is a new and growing body of research on learning and teaching in science that can . . . revitalize science education.”[4] Several pages later, it is stated that: “The framework is based on a rich and growing body of research on teaching and learning in science . . .”[5] From this it could be anticipated that much of the input for the *Framework* was provided by educationalists, a term I use for specialists in the theory and methods of education, and in the administration of education.

The *Framework* begins with the assertion that one of the failures of current science education in US schools is that it “emphasizes discrete facts with a focus on breadth over depth . . .”[6] The same point is made several pages later, with the oft-repeated quote that “science curricula in the United States tend to be ‘a mile wide and an inch deep.’”[7] The intent of the document not to repeat the same mistake is made abundantly clear: “the framework focuses on a limited number of core ideas in science and engineering. . . . The committee made this choice in order to avoid shallow coverage of a large number of topics and to allow more time for teachers and students to explore each idea in greater depth.”[8] While I am in complete agreement with this, the inclusion of engineering and technology within science is counterintuitive. How could engineering and technology be included *without* decreasing science content? The rationale for their inclusion is that

> Engineering and technology are featured alongside the natural sciences (physical sciences, life sciences, and earth and space sciences) for two critical reasons (1) to reflect the importance of understanding the human-built world and (2) to recognize the value of better integrating the teaching and learning of science, engineering, and technology.[9]

Neither reason strikes me as validation for adding engineering to the science curriculum, especially since engineering principles form integral parts of so many things that are studied under the rubric of science anyway. Indeed, it is difficult to visualize teaching science without teaching some aspects of engineering. How, for example, could a biology teacher do an adequate job of explaining the function of skeletons and bones without discussing beams, columns, trusses, forces, tension, compression, stiffness, and the like, all of which fall within the domain of engineering?

As far as technology is concerned, I see less opportunity for the inclusion of this subject matter within science. If I were teaching youngsters about air pressure, I would say something about Newcomen's atmospheric engine and, if bird flight were being treated, comparisons would be made to the structure of aircraft wings. However, these technological discussions would be very much subordinate to the physical and biological aspects being considered.

I can understand the rationale for wanting to teach engineering and technology to school students, on the grounds of increasing their chances of obtaining jobs in these areas, or preparing them for continuing into higher education. However, these disciplines should not be squeezed into the science curriculum, thereby diminishing its content. Instead, they should be taught as separate subjects. In similar fashion, technical and vocational subjects, like engineering drawing, metalwork and woodwork, shorthand and typing, and car maintenance, were taught in many UK schools during the fifties and sixties, along with more academic subjects like English, mathematics, history, and science.

Some would argue that teaching science, technology, and engineering as an integrated entity has greater value in that it cultivates the concept of their interconnectedness in the minds of students. Similar arguments could also be made for incorporating mathematics into science, too, as reflected in the promotion of STEM education. However, I cannot be convinced that students using mathematics in the science lab that was learned in a math classroom are in any way disadvantaged over those who were taught mathematics in science. Given that writing lab reports has as much to do with conducting science as working out a mathematical equation in a dynamics experiment, one could make the equally specious argument that English should be incorporated into science, alongside mathematics. Educationalists, with their fondness for acronyms, could then champion STEEM education; the spelling is atrocious but that matters little in the texting generation.

Before leaving the subject of spreading science too thinly, I should point out that the *Framework* document includes earth and space sciences, alongside the physical sciences (physics and chemistry) and life sciences. It is difficult to see how learning about such things as light from other galaxies, or volcanism, can be accommodated in the science classroom, along with everything else, while maintaining the intent of the *Framework* "to avoid shallow coverage." It is equally difficult to visualize how teachers could cope, either academically or practically, with such a wide-ranging curriculum.

At the last high school where I was a teacher, I taught biology throughout the school. My timetable was quite full, and I found it challenging enough just keeping on top of everything in the biology classroom. Even if I could have squeezed in some more teaching periods during the week, my science background did not extend beyond biology. Perhaps things have changed since those times, but this seems unlikely. Are there *really* high-school teachers today who could competently instruct students in everything from plate tectonics and natural selection to the big bang theory? Will schools start hiring new teachers with degrees in astronomy, geology, and astrophysics, along with those qualified in engineering and technology? Likely not, I suspect.

The *Framework* is "built around three major dimensions. . . . These dimensions are

- Scientific and engineering practices
- Crosscutting concepts that unify the study of science and engineering through their common application across fields
- Core ideas in four disciplinary areas: physical sciences; life sciences; earth and space sciences; and engineering, technology, and applications of science."[10]

Taking the first of these three "dimensions," I assumed that *scientific practice* pertains to the scientific method, like making observations, conducting experiments, and the like, but I had to be sure. The term, which occurs throughout the *Framework*, is used over thirty times before being defined, so I had to resort to the index to find it: "We use the term 'practices' instead of a term such as 'skills' to emphasize that engaging in scientific investigation requires not only skill but also knowledge that is specific to each practice . . . part of our intent in articulating the practices in Dimension 1 is to better specify what is meant by inquiry in science and the range of cognitive, social, and physical practices that it requires."[11]

Little the wiser, I returned to where I had left off. On turning the page, I found a table listing a number of attributes for each of the three dimensions of the *Framework*.[12] The eight scientific practices ranged from "asking questions" to "engaging in argument from evidence." While this confirmed that my original assumption had been correct, I still had a problem with the second dimension, the crosscutting concepts.

I think of *concepts* as ideas that can explain observable facts, like the concept of evolution explaining the changes seen in the fossil record over long periods of time, or the concept of plate tectonics explaining the way continents are slowly drifting apart. However, the seven items listed as crosscutting concepts include such things as "patterns," "energy and matter," and "scale, proportion, and quantity."[13] While these items occur repeatedly across different areas of science, they could hardly be described as *concepts* and therefore have limited value in unifying the study of science across different fields, as was the original intent.

The third dimension of the *Framework*, "core ideas," contains a series of topics to be studied in the various disciplines. Ranging from "energy" to "biological evolution," this seems quite straightforward. However, I do have problems with some of the items, including "waves and their applications in technologies for information transfer" and "links among engineering, technology, science, and society."[14]

In reading through the first couple of dozen pages of the document, I found several areas of accord with my own views, including the need for students to carry out scientific investigations themselves. Indeed, the document is dotted with photographs of youngsters of all ages involved in hands-on activities. The point is also made that youngsters, even of kindergarten age, have surprisingly sophisticated ways of thinking about the world, partly because of their own experiences in observing and exploring their environment. My experiences as a teacher opened my eyes to the remarkable learning abilities of the young, but it was only when I had grandchildren that I discovered how early this begins.

One of my grandsons, who is especially observant, is remarkably intuitive. Some years ago, while his father was assembling some furniture at home, his son pointed out that he had forgotten to include one of the parts—he had seen this by looking at the diagram. He was only three at the time.

Another aspect of the *Framework* that resonates with me is building upon a student's knowledge. The plan is to present given topics at each grade level so that students can build upon their prior understanding of them. Maintaining such a progression of learning throughout the student's school years is sound teaching practice.

Connecting with students in the classroom is another objective. Referring to several publications in the educational literature, the point is made that a child's "personal interest, experience, and enthusiasm"[15] is critical to their learning of science in school. To sustain that interest in science, the "classroom learning experiences in science need to connect with their own interests and experiences."[16] To that end, one of the four criteria used by the *Framework* committee to decide upon core ideas is to: "Relate to the interests and life experiences of students or be connected to societal or personal concerns that require scientific or technical knowledge."[17]

Attracting students to science by teaching them about things that interest them personally may be the ideal approach from an educationalist's perspective, but it strikes me as otherwise from a teacher's standpoint. Lesson content should be determined by educational merit—that a particular aspect is important to learn about and to understand—not by its personal appeal. Besides, how would students know whether something they had never experienced might be of interest to them? I doubt that any of my eleven-year-old classmates in science had ever experienced anything to do with air pressure before, but when Mr. Jordan collapsed that gallon tin can we all sat up and took notice. And societal concerns should have no influence on curricular content, as discussed earlier (chapter 1).

One of the sections in the *Framework* discusses how educational studies have shown that children have surprisingly sophisticated ways of thinking about the world, based upon their own experiences. This happens at an early age, even before they start school: "By listening to and taking these ideas seriously, educators can build on what children already know and can do."[18] When I read that passage, I was reminded of something that happened to me at a local elementary school, just a few years ago.

The school, where two of my grandchildren were in attendance, was having a "scientists in the classroom" morning, and I was one of the participants. The young teacher who escorted me into the classroom where I would be stationed explained how, every half hour, a new group of children would replace the ones already there. While we awaited the first arrivals, I set up some things on a table at the front. Needless to say, I would be talking to the youngsters about dinosaurs.

In addition to Norman, a small and anatomically correct *Tyrannosaurus* skeleton I had built from chicken bones, I had brought along an assortment of dinosaur bones and bird bones, along with some other items of interest. I planned on showing and telling the youngsters about things they had likely never experienced before, but the teacher had other ideas. She wanted me to begin by inviting the children to tell me everything they knew about dinosaurs. Once they had shared their knowledge with me, I would tell them what I knew. I did not think this was such a good idea, as there would be insufficient time to cover all the things I had intended to do. She could see I was unimpressed, but I agreed to do it her way regardless.

After listening to all the usual things that children pick up from misinformed books and TV programs—*T. rex* was the fastest dinosaur in the world and could run at over 30 mph or whatever—it was almost time for the next class. As the first group began filing out, the teacher approached for a brief discussion. She said she thought it best if I did not ask the children what they knew about dinosaurs. Common sense had returned to the classroom.

In making the decision to cover a smaller number of core ideas, the *Framework* committee recognized that this would disappoint some scientists and teachers by leaving out some of their favorite topics. However, they were convinced that by building a strong base of core knowledge founded on depth, students would leave school with a better grounding in science. This was far better than covering "multiple disconnected pieces of information that are memorized and soon forgotten once the test is over."[19] While I am in complete agreement with the importance of building strong foundations in science, I still see benefits in committing some facts to memory. Rote learning is anathema to many educationalists, but I can make a good case for memorizing facts, as the following example illustrates.

A couple of years ago, I was watching an episode of *CSI*, the popular TV crime series. I had never seen the program before and wondered what I might be missing. In this particular episode, somebody was pushed from the top of a thirty-story building. In the investigation that followed, it was asked how long the person took to fall to the ground. The number of seconds given by the investigator did not sound right to me, so I did my own calculation. To do this, I used equations remembered from my school physics days. There are three equations linking time, distance, velocity, and acceleration, namely: $s = ut + \frac{1}{2}at^2$, $v^2 = u^2 + 2as$, and $v = u + at$. The four variables in the equations are: u = starting velocity (in this case zero); v = final velocity; a = acceleration, in this case the acceleration due to gravity (32 feet per second per second), and s = distance traveled (300 feet for a 30-story building). Inserting these values into the first equation gave the time taken to fall as 4.3 seconds. Out of interest, I used the other equations to determine the speed at impact. This worked out to be 138.6 feet per second, or 94 mph (the slowing down due to air drag can be ignored for such a short distance). The 4.3 seconds taken for the victim's fall was considerably different from that given by the investigating officer, and I lost interest in seeing any more episodes. I have many more facts committed to memory from school science days, which have been of immeasurable use over the years.

Learning and teaching science used to be simpler and far more straightforward than it is today. Some would argue this is because science and technology have become more complex, along with the rest of the world, and that teaching science has to change accordingly. I do not see things that way. Science to me is simple and straightforward. It is a practical subject that involves being a careful observer, learning from what is seen, thinking things through logically based on facts, and arriving at conclusions that can be tested experimentally. I can illustrate this with a simple example.

I often take winter walks with my grandchildren, usually on cold blue-sky days. We sometimes see fallen leaves in the snow, russet reminders of fall days past. These leaves are often sunken down in the snow, sometimes to a depth of several inches. Why is this? I asked my grandchildren. Maybe some animal, like a rabbit, tried eating it and inadvertently pushed the leaf down into the snow. If that were so, why are there no animal tracks?

Perhaps the leaf fell when the snow was light and fluffy, imprinting itself like a footprint. But could a leaf fall with sufficient force to do that? When I asked them what the obvious difference in appearance was between the leaf and the snow, they said the leaf was much darker. I explained how dark surfaces absorb heat more readily than light ones, something remembered from school physics. This is why light-colored clothes are cooler in the summer sun than dark ones. This could be tested with a simple experiment with two rectangles of white card, about two inches (5 cm) long. One is blackened with a magic marker, while the other is left unchanged.[20] Both are then placed, side by side, on the snow in bright sunshine. After two or three hours, the black piece will have receded into the snow.

After the introductory and explanatory material of the first two chapters, the third chapter of the *Framework* is devoted to discussing how science works and explaining and rationalizing the changes taking place in the teaching of the subject.

"The idea of science as a set of practices has emerged from the work of historians, philosophers, psychologists, and sociologists over the past 60 years," begins a segment titled "Understanding How Scientists Work."

> Seeing science as a set of practices shows that theory development, reasoning, and testing are components of a larger ensemble of activities that includes networks of participants and institutions [10, 11], specialized ways of talking and writing [12], the development of models to represent systems or phenomena [13–15], the making of predictive inferences, construction of appropriate instrumentation, and testing of hypotheses by experiment or observation [16].[21]

Nine of the ten authors in the references cited in the square brackets are psychologists, and all are interested in childhood education. Delving into some of these works, along with those cited later in the chapter, was an illuminating, albeit frustrating, experience. One point to emerge was that most of what educationalists say about how scientists work is contrary to my own experiences as a scientist. It also became apparent why science education has become so inordinately complex. Mindful that the next few pages are not easy reading, it is necessary to explore these writings to get an idea of the way educationalists view science and how it should be taught.

One of the papers cited above [15], written by Drs. Richard Lehrer and Leona Schauble, both educational psychologists, explains that

> [The] scientists' work involves building and refining models of the world. Scientific ideas derive their power from the models that instantiate them, and theories change as a result of efforts to invent, revise, and stage competitions among models. These efforts are mobilized to support socially grounded arguments about the nature of physical reality, so model-based reasoning is embedded within a wider world that includes networks of participants and institutions; specialized ways of talking and writing; development of representations that render phenomena accessible, visualizable, and transportable; and efforts to manage material contingency, because no model specifies instrumentation and measurement in sufficient detail to prescribe practice.[22]

I must confess to understanding very little of this convoluted text, including the reference to competitions that we scientists apparently stage for our models.

As the paper draws to a close, we are told of one of the perceived benefits of model-based reasoning to the student:

> From a modeling perspective, one important outcome of a student's school career over the long term should be a growing repertoire of models that are extensible, general, and mathematically powerful.[23]

Whatever all this means.

This paper appeared in *The Cambridge Handbook of the Learning Sciences*. The editor of this 648-page "handbook," Dr. Keith Sawyer, another psychologist, defines the term *learning sciences* of the book's title as "an interdisciplinary field that studies teaching and learning."[24] He goes on to say that

> The goal of the learning sciences is to better understand the cognitive and social processes that result in the most effective learning, and to use this knowledge to redesign classrooms and other learning environments so that people learn more deeply and more effectively. The sciences of learning include cognitive science, educational psychology, computer science, anthropology, sociology, information sciences, neurosciences, education, design studies, instructional design and other fields.[25]

Dr. Sawyer tells his readers that

> This handbook is your introduction to an exciting new approach to reforming education and schools, an approach that builds on the learning sciences to design new learning environments that help people learn more deeply and effectively.[26]

From a perusal of the contents, this volume appears to have been written by educationalists for educationalists, rather by scientists for science teachers. Without wading through this thick tome, I cannot be definitive. However, judging from the small sampling I have made, I predict that it is full of esoteric discussions of science and learning that would be as confusing to scientists as it would be to science teachers.

In the last paragraph of "Understanding How Scientists Work," the *Framework* explains that

> The focus here is on important practices, such as modeling, developing explanations, and engaging in critique and evaluation (argumentation), that have too often been underemphasized in the context of science education. In particular, we stress that critique is an essential element both for building new knowledge in general and for the learning of science in particular [19, 20]. Traditionally, K–12 science education has paid little attention to the role of critique in science.[27]

Dr. Ford, an educational psychologist and author of the first paper cited above [19] tells us that

> It is clearly important for students to understand something about the architecture of scientific knowledge. In science as a social practice, critique motivates authentic construction of knowledge that is uniquely scientific.[28]

And, on the next page,

> A growing body of work in science studies asserts that disciplinary authority in science is social. That is, rather than its basis being in what an individual can do alone, it rests in the community of scientific peers. Members of this community work in a particular area and make decisions regarding what counts as a new knowledge claim.[29]

The notion that a community of scientific peers decides on what constitutes new knowledge is as fanciful as declaring science to be a social practice. Most of the scientists I knew, including myself, spent the majority of their time working alone on their research projects. Certainly there are joint projects, sometimes involving many researchers, as evidenced by the multi-authored publications that appear in scientific journals. The manuscripts submitted to journals for publication are also sent out for peer review by other specialists in the field—typically to one or maybe two reviewers, certainly not to a "community" of them. Scientists also attend meetings where they give presentations of their research, and this often receives feedback. Researchers also receive comments and criticisms of their work in papers published by their peers. However, this is all a far cry from the idea that a "community of scientific peers" working "in a particular area . . . make decisions regarding what counts as a new knowledge claim."

In the second cited paper [20], Drs. Leema Berland and Brian Reiser, specialists in learning sciences and in cognitive science respectively, tell us that

> Constructing scientific explanations and participating in argumentative discourse are seen as essential practices of scientific inquiry. In this paper, we identify three goals of engaging in these related scientific practices: (1) sensemaking, (2) articulating, and (3) persuading. We propose using these goals to understand student engagement with these practices, and to design instructional interventions to support students. Thus, we use this framework as a lens to investigate the question: What successes and challenges do students face as they engage in the scientific practices of explanation and argumentation? We study this in the context of a curriculum that provides students and teachers with an instructional framework for constructing and defending scientific explanations.[30]

Here again, I find myself wondering where non-scientists get their ideas about the way scientific inquiry proceeds. And why do they make science so complicated and so desperately dull for students and teachers alike?

The next segment of the *Framework* is headed "How the Practices Are Integrated into Both Inquiry and Design." This begins,

> One helpful way of understanding the practices of scientists and engineers is to frame them as work that is done in three spheres of activity, as shown in Figure 3-1. In one sphere, the dominant activity is investigation and empirical inquiry.
>
> In the second, the essence of work is the construction of explanations or designs using reasoning, creative thinking, and models. And in the third sphere, the ideas, such as the fit of models and explanations to evidence or the appropriateness of product designs, are analyzed, debated, and evaluated [21–23].[31]

The Figure 3-1 referred to above is a flowchart, reminiscent of the ones I saw in the Ontario science curriculum, with arrows linking boxes containing various words, labels, and catch phrases. I generally find flowcharts—widely used in everything from sales presentations to managerial meetings—to be as unhelpful as the message.

What is to be made of all this theorizing on the nature and practice of science? That depends on one's perspective. As the reader of this book, you may be wondering why you need to know about these writings of educationalists. You may also be wondering what it all means anyway. As a scientist, I find most of what has been written about what scientists do as ludicrous. I think I would fail a "being-a-scientist" test. For those with expertise in any of the areas represented by the authors I have quoted, I am sure they would feel engaged and on familiar ground. This would certainly be true for the two psychologists[32] who wrote the first of the three articles referred to in the last *Framework* excerpt [21], and also for the nine educationalists[33] who wrote the third article [23]. As an example of the esoteric nature of these two articles, I offer the following excerpt, written by the two psychologists:

> We start by summarizing the key features of our model of scientific discovery as dual search (SDDS). It is proposed as a general model of scientific reasoning that can be applied to any context in which hypotheses are proposed and data is [*sic*] collected. The fundamental assumption is that scientific reasoning requires search in two related problem spaces: the hypothesis space, consisting of the hypotheses generated during the discovery process, and the experiment space, consisting of all possible experiments that could be conducted. Search in the hypothesis space is guided both by prior knowledge and by experimental results. Search in

> the experiment space may be guided by the current hypothesis, and it may be used to generate information to formulate hypotheses.[34]

The subject of hypotheses and hypothesis testing is obviously of fundamental importance to those devising content and formulating policy for teaching school science. They likely think that scientists spend most of their time formulating and testing hypotheses, but I believe this overstates how much attention scientists pay to theorizing over science. Scientists are people with curiosity whose observations of what they see raises questions, and possible answers. Seeing a dead leaf partly sunk in the snow leads to the likely answer that it is because the dark leaf absorbs heat from the sun. Formally stated, the likely answer is the hypothesis, and this is tested by the simple experiment with light and dark objects placed on the snow. I suspect the thought processes of most scientists noticing such a leaf in the snow would be: "What's going on? Okay, the leaf's in the sun and it's dark in color so it's absorbing more heat than the surrounding snow. Let's check it out. Pieces of black and white card should do it . . ."

Educationalists spend too much time theorizing over hypotheses and models and all the other things scientists are supposed to spend so much of their time doing. To give a better picture of reality in the science lab, I will describe one of the research projects I worked on.

My specialty as a paleontologist was ichthyosaurs, a group of extinct reptiles that lived in the sea at the time dinosaurs roamed the land. The name ichthyosaur means fish-lizard in Greek and, if you saw one in the flesh, you might think it was a shark. Ichthyosaurs had paired fins, fore and aft, a large tail, a dorsal fin in the middle of the back, and most of them had a long pointed snout, armed with sharp teeth. The feature of interest in this particular study was the tail.

While the upper lobe of the tail had no internal skeleton, the lower lobe was supported by a downward kink of the vertebral column. This bend is formed by a change in shape of the centra, the bony discs that form the bulk of the vertebrae. While the centra are of regular disc-shape everywhere else, those at the base of the tail are wedge-shaped, being thicker at the top than the bottom. There are about six of these tapered centra, each making a small contribution to the bend. Measuring the angle of the tailbend gives an idea of the shape of the tail, which varies among the different species.

Figure 5.1: Some ichthyosaur skeletons from the early Jurassic locality of Holzmaden, in Germany, are so well preserved that their body outlines have been imprinted on the rock as a carbonaceous film.

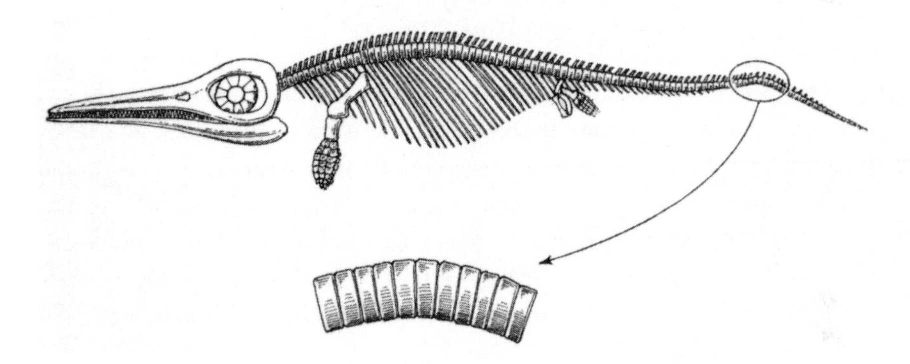

Figure 5.2: An ichthyosaur skeleton showing how the tailbend is formed by the slight tapering of the bony discs, called centra, at the base of the tail. These half dozen or so wedge-shaped centra are wider at the top than at the bottom.

The tailbend is an obvious feature in skeletons that lie on their side, provided the tail vertebrae have not been displaced, and measuring the angle is quite straightforward. Unfortunately, this is often not the situation and sometimes, as in the case of the species *Leptonectes tenuirostris*, there had been some uncertainty as to whether a tailbend was present. Resolving the issue was the objective of this particular study. For this, I had an almost complete skeleton of the species to study, borrowed from another institution.

This specimen, lying on its front with the fins splayed out at the sides, was contained in several slabs, mounted inside a wooden frame. The region of the vertebral column where a tailbend would be located was lying in the middle of one of the slabs, and this was easily removed from the rest.

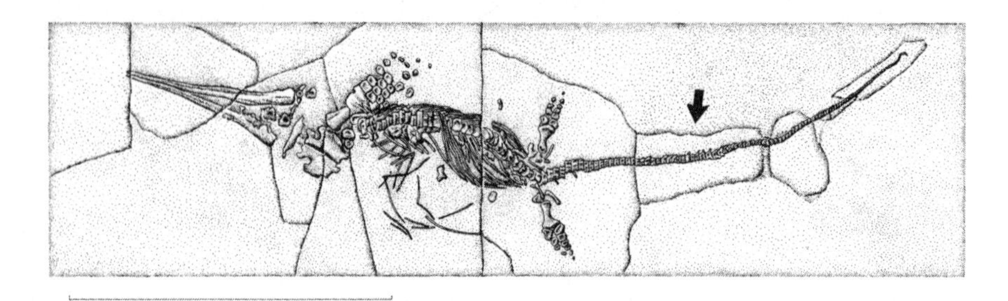

Figure 5.3: Drawing of the complete skeleton of the ichthyosaur, *Leptonectes tenuirostris*, that I used to determine whether a tailbend was present. The arrow marks the slab containing that part of the vertebral column where the tailbend would be found, if present.

One way to resolve the problem would have been to separate the bones from the rock and reassemble the tail. Aside from being a slow and laborious process, I did not have permission to do this. A noninvasive solution to the problem was to use CT scanning, a novel approach to investigating fossils back in 1989. I planned to generate a series of radiographic slices that passed vertically through the vertebrae. If there were any wedge-shaped centra, these would show up in the x-ray images, confirming that a tailbend was present.

From previous examinations of other specimens of *L. tenuirostris*, I anticipated that a tailbend would be confirmed. The next step was therefore to determine how to calculate the angle that each wedge-shaped centrum would contribute to the tailbend. Using simple geometry, I deduced an equation for this angle, based on three measurements: the height of the centrum and the thickness at the top and at the bottom.

Incidentally, my knowledge of geometry was acquired during mathematics classes at school, not during science.

The CT scanning revealed that six of the vertebrae were wedge-shaped, confirming that a tailbend was present. Their angles ranged between about 2° and 5°, giving an estimated tailbend angle of 24°; *L. tenuirostris* did not have a steeply down-turned tail. This research was published in the prestigious journal *Paleobiology*,[35] and was even featured by the late paleontologist and evolutionary biologist Stephen Jay Gould from Harvard University, in one of his regular articles in the magazine *Natural History*. My reason for raising this is not conceit but to make the point that I was a real scientist conducting real scientific research that was reported in a real journal of science.

Where was my hypothesis? That thought, I confess, had not crossed my mind, I simply wanted to know whether this particular ichthyosaur species had a tailbend and, if so, what was the angle of the bend. A specialist in science education would identify the hypothesis as: "*Leptonectes tenuirostris* has a tailbend," and this was tested using computed tomography. Rephrasing in terminology more appropriate to educationalists: "Having contextualized, in hypothesis space, during the discovery process, I formulated the hypothesis that *Leptonectes tenuirostris* has a tailbend. This necessitated formulation in experimental space, where an investigation was devised to test the hypothesis using computerized axial tomography scanning."

And what about models? The *Framework* points out that developing "models to represent systems or phenomena"[36] is a part of how scientists work. This point is made more forcefully in one of the papers cited in the *Framework*: "Modeling is a core practice in science and a central part of scientific literacy."[37] Where was my model? The authors of the cited statement have an illustration depicting a flashlight shining down on a solid block to create a shadow. This is described as an "Example model of how shadows occur." According to this, my figure deriving the angle of a wedge-shaped centrum would qualify as a model. Furthermore, it was "general and mathematically powerful."[38] However, I did not "stage competitions"[39] for my model, but I am a scientist not an educational psychologist.

While making models was obviously not an important part of my research, I recognize its importance elsewhere in science. In genomic research, for example, where large DNA data sets are involved, generating computer models to determine which branching patterns of relationships among the organisms are the best fit to the data is the only way to proceed. My reason for raising the issue of models is because I think too much emphasis is paid to this in the *Framework*, along with all the attention to hypothesis testing. The focus should be shifted away from theorizing over intangibles like mathematical models to conducting hands-on investigations using real things. No doubt conceptual models could be generated for the flow of air around objects of different shape to show which created the greatest drag force. But why would anyone choose to do this when students can make their own Plasticine models and test them with a hairdryer (chapter 4)?

Before leaving the subject of models, I would like to make a distinction between what I think of as real models and those that are abstractions. The former are objects that can be handled, like a plastic model of a DNA molecule. The latter include drawings, flowcharts, mathematical relationships, hypotheses, computer simulations, mental models, and the like, which are discussed and referred to throughout the *Framework*. While I find real models to be useful, like the cross sections of an ichthyosaur fin that I made from the cast of an actual specimen, I find abstractions less helpful. I have similar reservations about the use of abstractions in the science classroom, especially when real and functional models can be used. Who, for example, would want to use a computer simulation of an airfoil, freely available on the Internet, when you can build and test-fly your own, cut out from a cardboard tube? Clearly the *Framework* would want to do the former:

> Curricula will need to stress the role of models explicitly and provide students with modeling tools (e.g., Model-It, agent-based modeling such as NetLogo, spreadsheet models), so that students come to value this core practice and develop a level of facility in constructing and applying appropriate models.[40]

Notwithstanding that youngsters already spend too much time in front of screen devices, computer simulations are a poor substitute for hands-on experience. Take, for example, the trend in recent years for medical schools to discontinue the dissection of cadavers. Instead, students must resort to imaging technologies to "learn" about human anatomy. I certainly have no desire to receive any surgical treatment from a graduate of one of those medical schools. Nor do I want to board an aircraft flown by a pilot whose only previous training has been on flight simulators.

As already noted, another aspect of how scientists work, according to the *Framework*, "includes networks of participants and institutions . . ."[41] This, incidentally, repeats the exact same words used in one of the cited papers.[42] Participating in "argumentative discourse"[43] with other scientists is also part of the process. My research on the tailbend was clearly wanting here because I worked alone and did not enter into a single argument.

Before leaving the myths and realities of how scientists work, it would be useful to see how the *Framework*'s perception of science as a set of practices (quoted earlier in this chapter), is seen to compare with earlier perceptions of science:

> Our view is that this perspective is an improvement over previous approaches in several ways. First, it minimizes the tendency to reduce scientific practice to a single set of procedures. . . . This tendency overemphasizes experimental investigation at the expense of other practices, such as modeling, critique, and communication.[44]

From this it is clear that modeling, critique, communication, and all the rest, should not take second place to experimental investigations. The preference in school classrooms would therefore be for youngsters to learn about the "practices" of science, rather than focusing upon experiencing *real* science for themselves by conducting hands-on investigations.

The eight practices set out in the *Framework* are considered as essential elements of K–12 science and engineering. Each practice is discussed in some depth and the goals that students are expected to achieve by the end of grade twelve are given, along with guides to competence levels throughout earlier grades. Some expectations strike me as unrealistically advanced for school students, like being able to "Plan experimental or field-research procedures, identifying relevant independent and dependent variables and, when appropriate, the need for controls."[45] Based on my experiences of teaching undergraduate science students, most of them would be unable to achieve these goals, even by their final year at university. Nor would I expect undergraduates "to consider how to incorporate measurement error in analyzing data"[46] or to read "complex texts and a wide range of text materials, such as technical reports or scientific literature."[47] And I doubt whether many undergraduates would be expected to use "statistics to analyze features of data such as covariation."[48]

Aside from being unrealistic, I find some expectations to be incomprehensible. Consider, for example: "As they [students] become more adept at arguing and critiquing, they should be introduced to the language needed to talk about argument, such as claim, reason, data, etc."[49] The next sentence reads, "Exploration of historical episodes in science can provide opportunities for students to identify the ideas, evidence, and arguments of professional scientists." Even if this were true, is there ever time to indulge in such esoteric exercises in a science classroom? Nobody who has ever taught in one would think so. The *Framework* is replete with such assertions, along with tortuous passages of the kind favored by educationalists. Ironically, the *Framework* criticizes scientists for not writing clearly:

> Even when students have developed grade-level-appropriate reading skills, reading in science is often challenging to students for three reasons. First, the jargon of science texts is essentially unfamiliar; together with their often extensive use of, for example, the passive voice and complex sentence structure, many find these texts inaccessible [37].[50]

The publication cited above [37] appears in *Reading Science*, a book of academic papers, primarily written by linguists and educationalists.[51] In the opening chapter, written by the senior editor, it is stated that

> This book is by no means the last word as far as theoretical and descriptive recontextualisation is concerned. But it does demonstrate the productivity of opening up dialogue—across a range of socially and linguistically informed theoretical perspectives, across a range of institutional sites where science discourse is practiced, and across the language and image modalities through which science discourse is construed.[52]

This is hardly a paragon of lucid writing.

At the other end of the spectrum of expectations from those I consider too advanced for school students are those that state the glaringly obvious. Included here are: "Construct drawings or diagrams as representations of events or systems—for example, draw a picture of an insect"[53] and "use words, tables, diagrams, and graphs."[54]

Sometimes the obvious is stated in the language of educationists:

> Any education in science and engineering needs to develop students' ability to read and produce domain-specific text. As such, every science or engineering lesson is in part a language lesson, particularly reading and producing the genres of texts that are intrinsic to science and engineering.[55]

And sometimes the same obvious point is repeated elsewhere in the *Framework*:

> Students should write accounts of their work, using journals to record observations, thoughts, ideas, and models. They should be encouraged to create diagrams and to represent data and observations with plots and tables, as well as with written text, in these journals.[56]

A tendency throughout much of the *Framework* is to complicate the simple, as in,

> Students should be asked to use diagrams, maps, and other abstract models as tools that enable them to elaborate on their own ideas or findings and present them to others [15].[57]

Here, diagrams and maps have become "abstract models," which are then referred to as "tools." The cited publication [15] was written by the two educational psychologists, referred to earlier in the *Framework*.[58] The writing itself is often complex too: "As they engage in scientific inquiry more deeply, they should begin to collect categorical or numerical data for presentation in forms that facilitate interpretation, such as tables and graphs"[59]—meaning that students should use tables and graphs.

When it comes to enigmatic writing, some parts of the *Framework* have me completely baffled, as in: "We use the term 'models' to refer to conceptual models rather than mental models."[60] I also have problems with some of the inordinately sweeping statements:

> Scientists and engineers investigate and observe the world with essentially two goals: (1) to systematically describe the world and (2) to develop and test theories and explanations of how the world works.[61]

A similar statement is made later on, with implications for how students should proceed:

> Because scientists achieve their own understanding by building theories and theory-based explanations with the aid of models and representations and by drawing on data and evidence, students should also develop some facility in constructing model- or evidence-based explanations.[62]

This gives the impression that formulating theories is a regular part of being a scientist. It also reiterates the importance of making models.

From my perspective, too much emphasis is placed on modeling, along with hypothesis-testing and all the other methodological aspects of science. Meanwhile, the practical side of science—conducting experiments and learning from hands-on experience—is largely overlooked. This focus is seen throughout of the *Framework*, as when,

> Older students should be asked to develop a hypothesis that predicts a particular and stable outcome and to explain their reasoning and justify their choice. By high school, any hypothesis should be based on a well-developed model or theory. In addition, students should be able to recognize that it is not always possible to control variables and that other methods can be used in such cases—for example, looking for correlations (with the understanding that correlations do not necessarily imply causality).[63]

I cannot think of a better way of turning youngsters off science than with this endless rhetoric about hypotheses, testing, conceptual models, and the like.

Before leaving the "practices" section of the *Framework*, I should point out two errors. The first pertains to the confusion between a hypothesis and a theory:

> Science begins with a question about a phenomenon, such as "Why is the sky blue?" . . . and seeks to develop theories that can provide explanatory answers . . ."[64]

What is being described here is a hypothesis, not a theory. Only when a hypothesis has been extensively tested without being falsified does it become a theory.

The second error concerns using the words *precision* and *accuracy* synonymously. In everyday usage they have the same meaning, but this is not true in science. Accuracy is the closeness of a measurement to its true value. Precision is the closeness of repeated measurements of the same feature.[65] To illustrate the point, suppose you were measuring the length of a bone with some calipers, where one of the ends was slightly bent.

You would be unable to measure the bone *accurately* because the calipers would always give false readings. However, you would be able to remeasure the bone and obtain the same length each time, showing your measurements were *precise*. Errors in the use of precision and accuracy occur in several places in the *Framework*, as in: "Decide how much data are needed to produce reliable measurements and consider any limitations on the precision of the data."[66] And again in: "Students continue to . . . cultivate their skills . . . in recognition of the need for precision in both the measurement and interpretation of data . . ."[67] In both cases it is the *accuracy* of measurements that is relevant, not the *precision* with which they are measured. Suppose students are conducting an experiment to measure the rate of evaporation of a certain volume of water from a Petri dish. Two weighing balances are available, one of which always under-reads the mass, by the same amount, while the other gives correct results. Students using the first balance will have precise results, but only those using the second balance will have accurate ones.

The next major section of the *Framework* is the crosscutting concepts. These were intended to bridge across different disciplines to give students an overall view of science. As mentioned earlier, one of the problems I have here is that the seven items listed are hardly concepts. For example, I have difficulty thinking of *patterns* as being used to explain any observable phenomenon. Contrast this with, say, the cell concept, which explains why organisms as diverse as elephants and mice are made up of similar microscopic parts. The absence of any predictability in the crosscutting "concepts" helps explain some of the trivialities, confusions, and absurdities that occur throughout this section. I could document numerous examples but in the interest of space a few must suffice.

The statement attached to *structure and function* explains that: "The way in which an object or living thing is shaped and its substructure determine many of its properties and functions."[68] This competes for pointlessness with: "Air has many properties that can be used for flight and for other purposes" mentioned in chapter 1 in connection with Ontario's science curriculum.

If I had to choose the most nonsensical statement in the *Framework*, the following would be a leading contender:

> Although any real system smaller than the entire universe interacts with and is dependent on other (external) systems, it is often useful to conceptually isolate a single system for study. To do this, scientists and engineers imagine an artificial boundary between the system in question and everything else. They then examine the system in detail while treating the effects of things outside the boundary as either forces acting on the system or flows of matter and energy across it—for example, the gravitational force due to Earth on a book lying on a table or the carbon dioxide expelled by an organism. Consideration of flows into and out of the system is a crucial element of system design. In the laboratory or even in field research, the extent to which a system under study can be physically isolated or external conditions controlled is an important element of the design of an investigation and interpretation of results.[69]

Aside from my difficulty in considering something like a book lying on a table as a *system*, I have trouble accepting that *scientists* imagine artificial boundaries around them, with all those flows of matter and energy. I am reminded of the pseudoscientific claims I heard from practitioners of alternative medicine about energy flows in and out of the human body. Thinking in terms of systems belongs in the realm of educationalists, not scientists.

And the following would belong in a "What on earth has this got to do with science?" category:

> Students should also be asked to create plans—for example, to draw or write a set of instructions for building something—that another child can follow. Such experiences help them develop the concept of a model of a system and realize the importance of representing one's ideas so that others can understand and use them.[70]

The problems touched upon here are exacerbated by the compulsion to be all-encompassing, an imperative seen elsewhere in the *Framework*. This is reflected in the breadth of most of the crosscutting concepts. The resulting confusion is made even worse by errors in the science.

One of the most glaring examples of scientific ignorance occurs during the extended discussion of the last crosscutting concept, *stability and change*. In the lead-up to the erroneous passage, the reader is told that stability is a balance of competing effects, but that students typically think of equilibrium as being a static situation. Accordingly, they misinterpret a lack of change in the system as evidence that nothing is happening. They need guidance, we are told, so they can appreciate that stability can be the result of many opposing forces. The point is then made that

> They should be taught to identify the invisible forces—to appreciate the dynamic equilibrium—in a seemingly static situation, even as simple as a book lying on a table.[71]

The theme continues on the following page with the point that "An understanding of dynamic equilibrium is crucial to understanding the major issues of any complex system." The stage is set for the erroneous passage,

> For example, the stability of the book lying on the table depends on the fact that minute distortions of the table caused by the book's downward push on the table in turn cause changes in the positions of the table's atoms. These changes then alter the forces between those atoms, which lead to changes in the upward force on the book exerted by the table. The book continues to distort the table until the table's upward force on the book exactly balances the downward pull of gravity on the book. Place a heavy enough item on the table, however, and stability is not possible; the distortions of matter within the table continue to the macroscopic scale, and it collapses under the weight.[72]

The reason for the stability of the book has absolutely nothing to do with the downward force ("push" is not a scientific word) of the book changing the position of the *atoms* in the table. Instead, it has to do with the *structure* of the table opposing the downward force of gravity acting on the book. The reason for the confusion is because the author of the passage is thinking at the atomic level rather than at the structural level. A parallel example will help clarify the situation.

Suppose you had a weight-pan suspended from a fixed point by several feet of steel wire. If you added some weights, the pan would lower, very slightly, but the amount would probably be too small to measure with a ruler. The stretching of the wire is due to the force exerted by the weights pulling the atoms of iron further apart. The movements of the atoms are infinitesimally small, but there are so many trillions of them that the accumulative effect could be measured with dial calipers, if not with a ruler.[73]

If you now disconnected the wire and wound it tightly around a wooden rod, you could transform the wire into a spring. Repeating the previous experiment using the spring would result in a large lowering of the pan, easily measured with a ruler. This time you are exploring the properties of a structure (a spring), not the properties of a material (iron).[74]

The upward force acting on the book is due to the *structure* of the desk, not to any forces between the atoms or molecules of its *material*. This can be demonstrated by substituting a thin plank of wood for the desktop, supported at either end by small pile of books. The horizontal plank is now a beam, and placing a heavy book in its middle will cause it to bow downward. Provided the book is not too heavy, the beam will stop bending down before it reaches the desktop. The upward force on the book is due to the *strain energy* stored in the flexed plank—the same energy that is stored in a flexed bow prior to firing an arrow.[75]

The error in ascribing the force acting on the book to atomic forces may have been influenced by the author's primary mistake of believing the book-on-the-table scenario to be an example of *dynamic* equilibrium. It is, of course, an example of *static* equilibrium, as the name implies. Having made that mistake, the only perceivable source of movement in the system is that of the atoms themselves. The end result of all this is the transformation of a simple system into something complex—and factually incorrect. This glaring example of scientific ignorance is difficult to understand in a document intended to provide a "set of expectations for students in science."[76]

On reaching the end of the crosscutting section (chapter 4 in the *Framework*), which marks almost the one-third point in the document, one arrives at the first of the four disciplinary core ideas sections. The subject areas covered here are physical sciences; life sciences; earth and space sciences; and engineering, technology, and applications of science (chapters 5–8). I read the life sciences section, the area with which I am most familiar. Aside from a few careless errors in science, mostly in the first dozen pages or so, I was quite pleasantly surprised at what I found, especially after all the problems encountered earlier. Much of what I read made sense, and there were no glaring factual errors—different hands had obviously been at work here. However, when I finished that section and moved on to chapter 9, "Integrating the Three Dimensions," things seem to have regressed. This was evidenced by the following example of blatant scientific ignorance:

> An example of use of energy should include internal motion (e.g., heartbeat), external motion (self-propulsion, breathing), or maintenance of body temperature.[77]

Here, the imperative to use categories and to interconnect subject matter—what might be described as the micromanagement of science—has gone to the absurdity of distinguishing between "internal motion," used here for the contraction of muscles *inside* the body, and "external motion," for the contraction of muscles presumed to be *outside* the body. From my anatomical understanding, all muscles are internal, from those of the heart and diaphragm to those of the legs. And, even if there were such a thing as "internal" and "external" motion, why would anyone want to make the ridiculous decision to distinguish between them anyway?

How did so much time and effort, by so many people with such good intentions to improve science education in the United States, go so far astray? This is the view from my perspective, a scientist and a teacher who enjoys the challenge of making complex ideas clear and accessible. These conclusions were reached by reading the *Framework* document, along with some of the cited supporting publications. As we have already seen, these papers, written by educationalists not scientists, have clearly influenced the content of the *Framework*, and this is where the problem lies.

I could be suspected of being selective and citing only papers that were authored by educationalists, but consider the facts: Of the 432 references cited in the *Framework*, only *eight* pertain to science.[78] The bulk of the remaining 424 references relate to education (85 percent), along with psychology, sociology, philosophy, cognition, linguistics, politics, and anthropology.[79] One of the science sources cited is a book titled *Bad Science*, written by Ben Goldacre. A medical doctor and science writer, Goldacre shares my views about the nonsense that now passes as science. His words were obviously not heeded in the *Framework*.

It is clear to me that the overriding influence of educationalists and other non-scientists is the reason why science is so misunderstood in the *Framework* and why it is subordinate to teaching philosophies. Others will disagree. How could this be so, they will argue, when there were scientists on the *Framework* committee, including two Nobel laureates? First, the six scientists were in the minority in the eighteen-member committee, which included nine educationalists, two engineers, and one mathematician. Furthermore, five of the seven staff members, who worked closely with the committee, were educationalists. And anyone who has ever served on a committee, especially a large one like this, will know the role that consensus plays, and how decisions often resort to the lowest common denominator. Some support for my perspective of the failings of the *Framework* is provided by the steps leading up to it.

The forward to the *Framework* document states that it "builds on the strong foundations of previous studies."[80] Among the studies listed is *Benchmarks for Science Literacy*, developed by the American Association for the Advancement of Science (AAAS).[81] Chapter 13 of this document, "The Origin of Benchmarks," makes for some interesting reading, which, I believe, helps explain the mindset behind the *Framework*. The need was recognized, "to create a set of tools for educators to use in designing K-12 curricula . . . so as to obtain the desired science-literacy outcomes."[82] The question then arose as to who should undertake such a demanding task: "We decided that . . . school teachers and administrators, advised by education specialists and backed by scientists, would be most likely to develop intellectually sound curriculum models and other curriculum-design tools that would prove credible to other teachers."[83]

Six twenty-five-member teams were recruited in different parts of the country. Each team comprised five elementary, five middle-grade, and ten high-school teachers, along with one principal from each level and two curriculum specialists. The teachers "had taught the life and physical sciences, social studies, mathematics, technology, and also other disciplines."[84] It goes on to say that, "Consultants from around the country offered their expertise at annual summer conferences, where staff and teams met to advance mutual tasks."[85]

The point is then made that, "To foster continuity of ideas, the teams were asked to discern useful connections within and among the typically separate disciplines."[86] This may have been the inception of the *Framework*'s imperative to link topics together.

Team members met with "scientists, engineers, mathematicians, historians, architects, and physicians to learn about their current work and raise questions about its applications and relation to environmental and social issues."[87] Team members also attended "lectures and read widely, especially in the history and philosophy of science."[88] The inclusion of engineers and mathematicians, and the references to environmental and social issues, along with the philosophy of science, foreshadows the shape of things to come in the *Framework*.

Another important piece of information is that: "From the very first summer, the teams were introduced to the research of children's learning in science, mathematics, and technology through the work of several prominent researchers in the field."[89] The importance attached to the significance of these learning studies is clearly evident from the overwhelming influence of educationalists in the *Framework* document.

"Team members had to imagine what progress students could make toward each . . . goal, a process that came to be called mapping because it required groups to link more sophisticated ideas in later grades to the more primitive ones suitable in earlier years." [90] An illustration of one of these hand-drawn maps shows a flow diagram linking boxes labeled with such things as "water disappears into the air" and "visible things might be made of huge numbers of invisibly tiny pieces."[91] This again draws attention to the importance of linking things together. Making connections between related phenomena, like the curved wings of aircraft and of birds, makes perfectly good sense. However, problems arise when nonexistent connections are made, as between the force of gravity on a book and between the atoms of the table upon which it rests. People ignorant of science should not be writing about how it should be taught.

"The Origin of Benchmarks" notes that, as the process drew toward a close, "teams, consultants, and staff struggled to agree on a format that would do justice to the substance and thought, several issues arose. . . . Suffice to say that discussions were long and intense."[92] One can imagine similar discussions taking place during formulation of the *Framework*, with consensus prevailing over content.

When I started reading the *Framework*, I had no preconceived ideas of what I might find. The misinformation about how scientists work was unexpected, but the glaring factual errors in science came as a complete surprise, especially given the involvement of such an august body as the American Association for the Advancement of Science. This travesty came about because educationalists, people with little or no grasp of science, were allowed to decree what science is and how it should be taught.

Educationalists have taken control of science education in American schools as surely as they have in Canada. Their inordinate influence is somewhat reminiscent of the persuasive powers of the creation scientists, back in the eighties. The creationists' arguments were transparently thin and easily demolished by those with a sound understanding of science, but educationalists' views seem to go largely unchallenged. I think this is because so few people outside their field can be bothered trying to understand what on earth they are saying. I also think they are doing more damage to science in our schools than the creationists could ever have dreamed. And one can only imagine the pseudoscientific nonsense that curriculum designers will weave from the yarn spun on the *Framework*.

Teaching science in school was so much simpler in the sixties.

CHAPTER 6

TO SIR, WITH LOVE: TEACHING SCHOOL IN THE SIXTIES

My interest in paleontology began during my third and final year as an undergraduate in London, studying at Regent Street Polytechnic—the Poly.[1] In working toward a bachelor's degree in zoology, we studied the entire animal kingdom, though it was the vertebrates, particularly the long-extinct ones, that captivated me. I read books on vertebrate anatomy and paleontology like novels and purchased my own copy of the definitive work *Vertebrate Paleontology*. The author, Alfred Sherwood Romer, a Harvard professor and world authority, could write to engage neophytes like me as well as specialists. I had already devoured his popular paperback *Man and the Vertebrates* and used to write enthusiastic letters about paleontology to my fiancé.

Liz, whom I had met on holiday three years earlier, lived in Southampton, on the south coast. The eighty-mile distance between her home and mine seems inconsequential today, but to a penniless student and his telephone-operator girlfriend—diligently saving for our planned wedding the following summer—we felt worlds apart. We only saw each other once a month, on weekend visits to our respective homes.

During the spring of 1965, finals year, I began making inquiries about full-time doctoral positions. I first wrote to Professor D. M. S. Watson, a revered paleontologist at University College London. He had retired from teaching and was no longer accepting graduate students, but he suggested I contacted his colleague Dr. Kenneth Kermack. He in turn told me to contact him after my finals, when I had the results, which was not until August. Meanwhile, I started looking for a teaching position for September. This was my fallback position in the event that things did not work out and I had to pursue my doctorate as a part-time student.

If I had to teach, my first choice would be to obtain a position at an adult technical college, like the one I had attended for my GCE A-level qualifications in zoology and chemistry. New technical colleges were being built all over the country, and the ideal post would be in the southwest, preferably near the sea. Obtaining a position at a teacher-training college was equally attractive. School-teaching was at the bottom of my list; the idea of trying to teach a class of adolescents who had no interest in learning was unappealing.

Each week I scanned the *Times Educational Supplement*, the primary posting for all teaching jobs. In April I saw a promising opening at a technical college in Hendon, on the outskirts of London, and sent off an application. I had high hopes when I was invited to attend for an interview, but these were soon dashed on the day. The interviewers made it abundantly clear that a fresh graduate stood little chance of securing a job at a technical college without additional qualifications. Completely discouraged, I began thinking I would have to spend a year at a training college to obtain a teaching diploma. Liz, even more dispirited than me, did not want to face the prospect of having to postpone our wedding for another year while I continued as a student. The only alternative was to become a schoolteacher.

There was no shortage of school-teaching positions and, although I was at something of a disadvantage without a teacher-training certificate, this had not yet become a requirement. Ideally, I would have liked a job south of the Thames, in familiar territory, but instead I found myself searching throughout the greater London area. My first interview, in late June after my final exams, was at a girls' school in Brixton. Generally perceived as a rough part of London, the area had a bad reputation for crime. The interview went well, and when it was over the headmistress took me for a short tour of the school. As we chatted, she voiced her concerns for how a young man like me would manage with her boisterous girls. I had no misgivings, but she seemed genuinely troubled for my wellbeing and did not offer me the job.

The next interview was for a position at a school in Essex, north of the Thames, on the northeastern outskirts of London. The interview took place at the local educational offices and, judging from the activity inside and the number of worried looking people like me milling around, it appeared to be some sort of job fair. I did not get the advertised position, but, after the interview, I was approached by a kindly lady who was looking for a biology teacher for a school in Dagenham. The Robert Clack Technical School, she said, was a very good school, with nice pupils and a good staff.

Unlike most of the other senior schools of the day, it was co-educational, and she was sure I would like it. Her car was just outside, so we could drive over there right away—she seemed keen to see me recruited.

On arriving at the school, I was introduced to the deputy head, a business-like woman in her forties who took me on an impromptu tour. The building was modern, like the last school I had attended as a student, and was of about the same modest size. From what I could see, the pupils were well behaved and the staff friendly; I could see myself fitting right in. After a short interview I was offered the job, which I gratefully accepted. The position was conditional upon my graduating, but my finals had gone well and I was confident of graduating with a good degree, maybe even an upper second.[2]

The deputy head said that, if I wished, I could start teaching at the beginning of the following week. With only two weeks left before the summer holidays, the school year was winding down, so this would be a gentle way to break me in to the teaching profession. There would be no lessons to prepare, and most of my time would be spent sitting-in for other teachers, who were either away or occupied with other duties. I agreed that this was an excellent idea.

I do not think I missed very much by not going to teacher-training college. The headmaster, a down-to-earth man in his mid-fifties, gave me all the preparation I needed, in his office on my first day. "Don't take anything from the little buggers," he began, in his strong northern accent. He went on to say that, given the chance, even the good pupils would be disruptive if you allowed them to: "Misbehaving is more fun than doing work!" All I had to do was let them know who was boss, right from the start, and not let them get away with anything.

The most effective way of dealing with bad behavior was to take away their freedom. This could be done by keeping them in during their lunchtime, or by giving them a detention at the end of the day. School ended at four p.m. and teachers could detain offenders for as long as an hour if they wished. "Just make sure to follow through with your threats though," he warned, "otherwise you'll never be able to control them."

All of this brought back memories of my own schooldays, where the lives of some of our teachers were made a living hell. They were the ones who never punished us, regardless of all their threats. I was determined that the pupils would never get me on the run.

The final exam results were posted early in August, and I have a mental picture of nervously checking the notice board outside Senate House, the administrative center of the University of London. Finding the pass list for zoology among all the other subjects, I scanned down from first-class honors to second-class honors (upper division) without seeing my name. So I did not get the good degree I was hoping for.

My heart was really racing as I checked through the next list: second-class honors (lower division). Without receiving at least a lower second degree, my chances of being accepted as a doctoral student seemed remote. My eyes flitted down the list without seeing my name. Please . . . Then I saw it. A lower second was not what I was after, but it was not a bad degree, especially for an external student.[3]

Having received my results, I made an appointment to see Dr. Kermack. We got along well, and he offered me a full-time studentship to work on my doctorate. The government stipend for doctoral students at that time was £500, which was about half of what I would be earning as a new teacher. Combining this with Liz's earnings as a telephone operator would not give us enough to live on. I had no choice but to decline his offer.

The pupils at the Robert Clack School ranged in age from eleven to eighteen and some of the older ones barely looked younger than me or my new wife. They followed the same curricula as I had at school, taking their O-level GCEs (Ordinary-Level General Certificate of Education exams) at sixteen and their Advanced Levels at eighteen. I taught general science to the lower grades and A-level zoology to a small class of two, who were at the start of their last two years at school.

No matter how well prepared I was, those first forays into the classroom were still quite daunting. But nothing went wrong, and I soon learned to relax and enjoy myself. The students all knew there was a thin red line that they must not cross, and I never had any discipline problems. They were good students and it was a happy school, all of which worked in my favor.

One of the first lessons I learned was how quickly young minds absorb things. I made this discovery after discussing the structure of atoms with a class of eleven-year-olds. In describing how the negatively charged electrons orbiting around the positively charged nucleus had almost no mass, I casually referred to them as "little gritty bits." During a class test a week or two later, one of the boys wrote the exact same words in answering a question on atomic structure. This highlighted the need for absolute clarity when explaining things to youngsters.

One of the aspects of teaching that gave me most satisfaction was getting ideas across to students. I was teaching science the way I had been taught, with the teacher doing most of the talking and showing, while the class attended and took notes—the chalk-and-talk method. One of the topics on the syllabus was osmosis, the movement of solvent molecules, usually water, across a semipermeable membrane separating solutions of different concentrations. Osmosis has considerable biological importance, being the mechanism by which animals absorb water from the gut and plants to take up water from the soil.

The basic mechanism was easy enough to explain. Water molecules were small enough to pass through the pores in membranes but large ones were not. Water molecules would therefore pass from the weakest solution, where they were most abundant, to the strongest solution, where they were fewest. Students would soon get the idea, but how could I demonstrate this with a simple experiment? Then I had an idea.

The white membrane beneath an eggshell was surely semipermeable. What if I dissolved the shell from a raw egg with acid and then immersed the shell-less egg inside a strong sugar solution? There would then be relatively more water molecules inside the egg than in the syrupy solution, so the water molecules would pass out through the membrane, and the entire egg would become flaccid. If the egg was then placed in water, the reverse would be true, and the egg would swell up and become turgid. After conducting the experiment at home and confirming everything worked as predicted, I was ready to demonstrate this to the students. After placing the swollen egg into a small jar of water for transporting it to school the next day, I topped up the water level and screwed on the top.

The experiment was enthusiastically received by the class, and youngsters crowded around the table at the front for a chance to bounce the rubbery egg themselves.

They were fascinated to see how eggshells could be dissolved with acid, and I told them they could repeat this at home by leaving an egg in vinegar for a day or so.[4] After explaining the mechanism of how the semipermeable membrane acted something like a sieve, I placed the plump egg into a beaker of the syrupy solution. Saying that it would take a few hours to see any appreciable change, I told them they could stop by at the end of the day to check if they wanted. Lots of keen youngsters returned to the classroom after school, intrigued to handle the flaccid egg, which had lost all its bounce.

Teaching at the Robert Clack School was an enjoyable experience, and I quickly adapted to my new profession. John Whittaker, who taught physics, was a new teacher too, and we became good friends. We shared the same table in the staff room and had a similar sense of humor.

I did not pick up on the nuances in the staff room as readily as I should have. Perhaps it was because of this that my suspicions were not aroused when one of the first staff members to introduce himself to me turned out to be the physical education teacher. Several days later he casually asked if I would referee a soccer match for him one afternoon after school. I was so taken off guard that I replied in the affirmative. The afternoon of the match duly arrived.

Devoid of any sporting genes, my knowledge of the rules of soccer was negligible. However, I thought my position of authority would allow me to bluff my way through with the boys on the playing field. Blowing my whistle each time the ball went over the sideline was easy enough, and I could spot when a player touched the ball with his hand. But then disputes began arising where opposing players turned to me for a ruling. "Offside!" was the most common appeal and, since I had no idea what being offside involved, I had only a fifty percent chance of making the correct call. "Foul!" was another tricky infraction, unless I saw something obvious, like a deliberate push. The players soon realized I knew nothing about their beloved sport, and it was only the thin red line between them and me that kept any order on the field. During one particular dispute between opponents I decided to be proactive, so I blew my whistle and shouted "Offside!"

"Offside, sir?" queried one of the players respectfully. "Why offside, sir?"

"Because we haven't had an offside for some time."

The following day a boy approached me on the street as I was walking toward the bus stop. He had missed the game but had obviously heard all about it from his classmates. "Please sir, next time you referee a match will you let me know?" he asked with a grin. "I wouldn't want to miss it for anything."

I was never asked to referee another match.

Although there were no discipline problems, I did have trouble with one particular class. Thesc boisterous girls were in their last year at school and some would likely be pushing baby carriages around the Dagenham estate within a few years. I was charged with teaching them some "useful household science," but nothing I did seemed of the slightest interest or use to them. In the movie *To Sir With Love*, a rookie teacher, played by Sidney Poitier, won over a similarly unruly final-year class by tailoring his teaching to their particular interests. This idealistic approach was not my way of teaching and, when the film was released, two years later, I recall remarking to Liz that this was not the way things worked in my experience of the classroom.

At one point during a particularly disruptive class, I quipped that the likely reason they were being so difficult was that they had a crush on me. This was greeted with howls of derision and a comment that nobody in their right mind would want anything to do with me. I was mindfully aware of everything I said to those young ladies after that careless comment, and some semblance of order was maintained.

While most of the pupils spent their lunchtimes in the playground, the older ones had the option of listening and dancing to records in the gym. We teachers therefore had an occasional gym duty interspersed with our playground patrols. Many of the older teachers found this irksome, but the youngsters were playing the same music that I listened to so I was quite happy.

The youngest of the new teachers, a petite and attractive blonde named Janet Box, had been a pupil at the school herself just three years earlier. She wore short skirts and high heels and must have driven the adolescent boys into hormonal overdrive. She taught domestic science, which was only taken by the girls, and it was probably just as well there were no boys in her classes. Her classroom was equipped with stoves, fridges, food-mixers, and a full range of kitchen utensils. Janet always seemed to be in a hurry, especially first thing in the morning, and never seemed to have time for breakfast before leaving home. Instead, she made coffee and toast in her classroom, and I sometimes used to join her for a coffee, often accompanied by John Whittaker.

One of the perks of being at the Robert Clack was that they taught auto mechanics and had a fully equipped workshop, complete with an inspection pit. Some of the teachers tinkered with their cars there on weekends, and I did the same when we bought our first car. This was purchased in the fall of '65, a 1947 Austin Ten that cost us £20 (equivalent today to about

$375 US). The most important event at that time, however, was the start of my doctoral quest. This all began with a letter, the happenstance of which underscores the enormous part that chance plays in our lives.

In a lecture on paleontology during my last year as an undergraduate, our lecturer happened to mention the name John Attridge. Attridge worked on dinosaurs, taught at Birkbeck College, in London, and was described as being very approachable. Now that I was settled in my new job I decided to find out more. I discovered that Birkbeck was unique among the University of London's colleges in being solely for part-time students—they could have had me in mind. After checking that there was a John Attridge on staff, I wrote him a letter explaining my circumstances and my wish to pursue a PhD in paleontology. He promptly replied that he would be happy to meet with me and that I should phone him at his office to arrange a time.

Birkbeck College is located in the Bloomsbury region of London, a fashionable area of small parks and colleges. Consulting my London pocket atlas a few days before the early evening appointment, I saw that it was likely less than an hour's subway journey from the school. Always concerned about punctuality, I allowed myself considerably more time than this.

Arriving at Tottenham Court underground station with time to spare, I had a leisurely stroll to the college. On inquiring inside for directions to Dr. John Attridge's office, I was directed to the second floor. Making my way along the corridor, passing lecture rooms and teaching labs, I found his small office around the corner on the right. The door was open and he was sitting at his desk. Standing up on seeing me, John strode across, smiling, and extended a hand in greeting. It was a huge hand, befitting the person towering above me. Well over six-feet tall and correspondingly broad, the charismatic giant immediately put me at my ease. Soon we were sitting and chatting as if we had always known each other. John was as interested in my unorthodox education as he was in my enthusiasm for paleontology.

From my perspective, the interview went really well. Only much later on did John confess to his consternation over one of my comments. In answer to his question on what group of vertebrates I was interested in studying, I had replied, "I wouldn't mind giving the old fossil dicky-birds a go." Regardless, John agreed to take me on as his doctoral student.

The first priority was finding a suitable research topic for me to work upon. To this end, John suggested we visit his colleague Dr. Alan Charig, Curator of Fossil Amphibians, Reptiles, and Birds at London's Natural

History Museum. Housing the largest collection of fossils in the country, this was one of the world's premier museums.

Arriving at the museum several days later, we made our way through the entrance hall, passing the skeleton of *Diplodocus* on the left, and turned right into the gallery of fossil marine reptiles. Walking the length of that gallery, with its floor-to-ceiling skeletons of ichthyosaurs and plesiosaurs, brought us to the door marked *Enquiries*. John and I both knew the way—I had visited there during my Poly days and been generously shown around the department by Cyril Walker, Dr. Charig's assistant.

After introducing me to the curator, John left us alone to talk. Once we had discussed my undergraduate experiences and interests, we got down to the question of a suitable research project. He identified two vertebrate groups that were in need of attention, both represented by large collections of specimens. The choice between Pleistocene pigs and ichthyosaurs was an easy one to make. Having decided on ichthyosaurs, he suggested I go out into the gallery to look at some of the skeletons. Most of the ichthyosaurs found in Britain were Liassic in age—the earliest part of the Jurassic—and these were the ones I would be studying. One of the primary collecting localities was Lyme Regis, where Mary Anning collected the first ichthyosaur, over two centuries ago, when she was only eleven or twelve.[5]

Aside from what I had read in Romer, I knew little about ichthyosaurs, so John lent me his copy of an 1881 tome that contained a wealth of information about them. This had been written by Richard Owen, an outstanding anatomist and one of the leading paleontological lights of his time. It was Owen who coined the name *dinosaur*. In addition to studying all the Liassic specimens in the museum's collection, many of which were mounted on the wall in the gallery, I would be making a detailed three-dimensional study of the skull. The latter investigation would be based on some partial skulls that had been completely removed from the rock. This had been done by dissolving away the encasing limestone with acetic acid, the same weak acid found in vinegar. This relatively new way of preparing fossils had been developed in the museum's preparation lab. As I would be doing some of this fossil preparation myself, arrangements were made for me to spend two weeks in the lab working with Arthur Rixon, one of the pioneers of the process. This was scheduled to take place during the school Christmas vacation.

I learned so much about the handling and extrication of fossils from

Arthur Rixon, lessons that would serve me well in the years to come. The trial specimen I had to work on was an incomplete ichthyosaur skull of limited importance that Cyril Walker had located in the collection. Crushed fairly flat, it lay embedded in a triangular slab of limestone, about eighteen inches long and an inch thick. All that was visible, peeking from the smooth gray rock, were parts of the eye socket, snout, and most of the lower jaw.

The first step was to apply a protective coating of varnish to the exposed bones with a paintbrush. The bones were then marked up with indelible ink, so that a photographic record could be made for later use in relocating any bones that became separated from the rest. The bone was then re-varnished to seal the labeling. Before it was ready for the acid bath, the back and sides of the slab were enclosed in a protective casing. This was built up using liquid latex rubber with some filler, followed by a fiberglass and polyester resin outer shell. Once the resin had hardened, the specimen was suspended, bone-side down, in a small bath of acid, to dissolve away the limestone. The protective casing ensured that the limestone was only dissolved away from one side, maintaining the integrity of the specimen. It also provided a convenient means of handling the specimen, which required regular daily inspections.

The preparation process would take a few months to complete so, just before the vacation ended, I took the specimen home. Returning to school with the skull after Christmas, I set it up in the small room at the front of my classroom. That way I could monitor progress each day, sharing the fossil-preparation experience with my students.

As I had not attended teacher-training college, I was scheduled to be assessed by one of Her Majesty's Inspectors of Education. Although I knew in advance which day the inspection would take place, there was no indication of what time this would happen, so I did not know which class would be inspected. I do not recall being anxious—I was always well prepared for my classes and never had any discipline problems. But the thought crossed my mind of what would happen if I were found wanting in some way. Would I lose my job? Would that be the end of my being able to teach?

The inspector duly arrived and sat, in silence, at the back of the classroom. I tried to pretend he was not there. The pupils all knew something was going on, but I think most of them believed they were under the microscope, not me. The lesson ended at the mid-morning break, and I dismissed the class. However, one boy remained seated. He had asked a question during class, and I had told him to stay behind afterward so I could go over the material in more detail. He left several minutes later, satisfied with my lengthier explanation.

At this point, the inspector came to the front to talk with me. His only criticism was that I should not have spent so much time with the boy at the end of the class. He made the point that I had already done enough for him during class time, and that I needed my break away from the pupils.

During school term, I used to go to Birkbeck one or two evenings a week and all day on Saturdays. John was an exceptional supervisor, providing guidance, encouragement, academic challenge, and friendship. He was also remarkably accommodating, essentially turning his small office over to me. Soon my ichthyosaur material spread out to occupy most of the bench space next to the window, leaving him with just his desk and bookshelves.

Although Liz and I enjoyed our new life together, I never really settled down to living north of the Thames. We also had to contend with an interfering landlady. She lived in the lower half of her subdivided house and used to go snooping in our apartment when we were both at work. We decided that I would leave the Robert Clack at the end of the school year and find another school, south of the Thames. With a year's teaching experience to my credit, there should be no difficulty in finding a new position. I would certainly miss my first school, where I was very happy, but neither of us would miss our landlady. I do not recall when the decision to leave was made, but I do remember that it required handing in my notice sometime in May, before I had another job. We must have been more adventurous in those days, especially since Liz was expecting our first child.

Finding a new job was harder than I had expected, and there was surprisingly little choice. In the end I accepted a post at a secondary modern school in Bexleyheath, some ten miles from my hometown of Beckenham. The headmaster, large and powerfully built, could have been a rugby player in his youth. He came across as a no-nonsense man who wanted to provide a stable educational environment for his pupils, many of whom were below the intellectual norm for their age. This was obviously not going to be another Robert Clack, but I would manage; and it was located in the area where we wanted to live. Besides, September was a summer vacation away and there was much work to be done on the ichthyosaurs. We were also taking a holiday in the West Country, our favorite part of

England, where I would be meeting with some of the local fossil collectors of Lyme Regis and the neighboring village of Charmouth. I would also be searching the sea cliffs for fossils myself.

Bexleyheath School, built sometime before the Second World War, post-dated the Victorian era by generations, but the term Dickensian comes to mind when looking back. The building did not evoke that era, but the authoritarian rule that reigned there did. This was immediately apparent on my first day, when the school's head of science introduced me to my class, in the science lab. His short-cropped hair and towering presence lent a daunting demeanor, and he spoke with the "Cor Blimey!" accent of a Cockney.

As it was the start of the school day, the stools were still on top of the benches from the previous afternoon, left that way so the cleaners could mop the floor. With the senior teacher at the front of the lab, the boys obediently removed their stools and sat down. Leaving his stool on the bench, he began to harangue them about correct laboratory procedure, emphasizing the need for safety. While elaborating upon the perils of tomfoolery in the lab, he deliberately knocked over his stool. The unexpected crash made everyone jump. He then berated them on the dire consequences of breaking valuable equipment, like beakers and test tubes. By the time he had finished, every boy in the room looked too terrified to do anything. He then introduced me and left.

I spent most of the lesson trying to reassure and encourage the boys. My primary concern, I emphasized, was for them to learn about science. Accidents happened and, if a few test tubes got broken along the way, they would not get into any trouble from me, provided they were not being deliberately careless.

Although poles apart in our approaches to teaching, and with little in common besides our interest in science, the head of science and I got along famously well. He was a decade or more my senior and took an almost fatherly interest in my welfare at the school. He was also intrigued by my research, and I remember a discussion we had about how to resolve the problem of taking measurements of ichthyosaur skeletons that were mounted behind glass on the wall of the Natural History Museum. Measuring skeletons was going to be my next major task, scheduled for the school vacation the following summer.

Instead of engendering a spirit of cooperation and mutual respect among staff and students, the headmaster seemed to perpetrate a state of conflict. For him, it seemed, the boys were more akin to inmates at a correctional institution than pupils at a school, and the best way to maintain order was through intimidation rather than collaboration. The standard punishment for misbehavior was the cane, the free use of which was as much a part of school life as gun-slinging was part of the Wild West. This was to have unwelcome consequences for me during the establishment of the thin red line between me and my students.

New teachers are traditionally tested by their students to see where the line is drawn, and the outcome of those first critical weeks can determine the teacher's entire career at the school. If a teacher shows weakness in maintaining discipline, the children exploit it like a pack of wolves probing a herd of deer. And once they have their quarry on the run there is only one outcome. I remember one poor teacher when I was at school being reduced to a babbling wreck by a group of my classmates. Not content with hounding him in the classroom, they would follow him down the corridor, heckling and chanting their taunts. And every so often he would utter his familiar, "I'm warning you boys . . ." His tormentors could imitate the empty threat to perfection.

Without some form of punishment, it is almost impossible to maintain discipline, and I only ever encountered one teacher who could do this. Mr. Dyne was a quietly spoken, diminutive man who taught mathematics at my old high school. For reasons unknown, he maintained absolute control without even raising his voice—none of us dared step out of line.

The threat of receiving a detention after school was deterrent enough to maintain discipline at the Robert Clack, but this was utterly ineffective at Bexleyheath. The only deterrent that seemed to work was the cane. This was apparent during my first few days, when I was often asked if I was a hard-hitter. The thought of using a cane was abhorrent to me, but it became increasingly obvious that this was the only deterrent that would work. I was the new sheriff, and if I could not handle a six-shooter I would be run out of town. The role of Gary Cooper did not sit well with me, but I knew that High Noon was rapidly approaching.

The offender and his crime have long since been forgotten, but I still remember the caning. Retribution took place in the sick room, a dual purpose facility with a couch for the unwell and canes for the unwise. Punishment had to be witnessed by another teacher and the sentence duly logged in a book. When the fateful day arrived, I asked the head of science if he would witness the caning, set to take place after lunch. "Not arf!" was the enthusiastic positive reply.

I had no appetite that day, either for lunch or for what was about to happen. I was probably dreading it far more than the offender, for whom caning was a regular part of school life. Not only did I have to mete out the punishment, I also had to make sure it was effective. If the boy walked away bragging to his chums that I was not a hard hitter, I would be worse off than before.

At the appointed time, three people meet outside the sick room. The smallest is poised and prepared, the largest is relaxed and happy, the third is feigning all these things. While my witness makes an entry in the punishment book, I make an elaborate display of selecting a suitable cane. I pull out the first one, hold it up to one eye to check for straightness, make a few perfunctory swishes in the air, and then return it to the rack. The exercise is repeated for a second cane. From the corner of my eye, I can see my ploy is having the desired effect. Having made my final selection, I deliver a couple of trial blows to the couch, hitting it with all my might; I now have the boy's undivided attention. With a nod from me, my witness instructs the recipient to touch his toes.

When it was all over, the head of science admitted to being impressed by my performance. Meanwhile, the butt of my attention was spreading the word among his friends. The thin red line had been drawn.

Unfortunately, this was not the last caning I would have to deliver; corporal punishment was part of the culture at Bexleyheath. This was in marked contrast to the secondary modern school I had attended. During my two years as a pupil at Alexandra County Secondary School for Boys, I witnessed the cane being used but once. That was during a music lesson when the teacher was playing the violin. Some prankster at the back threw a handful of pennies down to the front. I thought it was quite funny, but Mr. Biggar did not think so. After delivering a series of strokes to the comedian's backside, the incident was forgotten.

The headmaster treated his staff much the same way as he did the boys. He decreed that teachers must hand in a notebook every Monday morning, outlining the lesson plans for the week ahead, with a review of the previous week's work. I showed my contempt by purposely forgetting to hand mine in. This used to infuriate him. On one occasion, he came into my classroom where I was teaching and stood by the door, beckoning. Pretending not to notice, I carried on with the lesson. Instead of taking the hint, he strode to the front of the class and said he wanted to see me in his office when I was finished. I later discovered that, far from discussing anything of importance, all he wanted to do was complain about my failure to hand in my lesson plans on time.

My job was to teach general science, along with some chemistry, a sprinkling of physics, and a fair amount of biology, all at a basic level. I do not recall any outstanding students, and few achieved more than a general understanding of science. At the other end of the spectrum, in what was designated the d-stream, was a class of problem pupils, a mixed group of low-achievers, many of whom were more interested in causing trouble and entertaining their friends than learning. A few of the boys had obvious intellectual disabilities and could barely function in the regular classroom situation.

One of the boys, in his early teens, used to entertain his classmates with a remarkably convincing imitation of an orangutan. He was usually quite manageable, but there was one day when he was so disruptive that I had no alternative but to order him from the classroom. I do not recall how this came to the headmaster's attention, but his response was immediate: the boy would report to his office at the end of the morning, where I would cane him—a command performance.

I arrived at the headmaster's office a little ahead of the offender. As soon as the boy appeared, he was admitted to the inner sanctum. And there he obediently stood while the headmaster opened the tall cupboard opposite his desk. This revealed a row of canes, neatly arranged like a collection of prized rifles. Pondering for a moment, he made his selection and handed me a cane that was easily as long as I was tall. Wielding such a cane with one hand, as was required, was extremely difficult, so I resorted to grasping it about one foot from the end.

At the headmaster's command the boy came forward, touched his toes, and I took aim. Controlling the cane was challenging, and the first stroke went too high, skittering ineffectively across the boy's back. The next one landed squarely on target, but the wayward end flexed around and caught him on the ankle. It is unclear whether the howl that emanated was due to the blow to his backside or to his ankle. Regardless, after being dismissed by the headmaster, the boy disappeared for the rest of the day.

A young teacher from South Africa started at the same time as me, and we used to spend much time together in the staff room, commiserating and exchanging classroom experiences. I recall showing him some of the answers I received from a test I had set for one of my classes. On a question about respiration, one of the boys had written, verbatim: "Breevin is very important becawse without it you could get ded very easily."

I did much of my teaching in a portable classroom in the playground. This was comfortable enough during the early autumn, but in November, when the gas heater was not working properly, it was frigid. Sometimes the caretaker could fix the temperamental unit. On other occasions, I dealt with the problem by having the boys light their Bunsen burners, which soon got the room cozily warm.

One day, the school was disrupted by a fire alarm and I had to assemble my class in the playground, along with all the other boys. The alarm turned out to be false and, regardless of the headmaster's best efforts, the culprit was never found. From the suspicious looks he gave me at the next staff meeting, I was half convinced he thought I had been responsible.

The usual morning greeting between me and my South African colleague, one that he had initiated, was "Another bloody day!" And that was exactly how it felt. I had to get away from that school, and I diligently searched the *Times Educational Supplement* every week. One job opening that caught my attention was for an assistant master of biology at Westminster City School, an independent grammar school for boys in the heart of London. The position was to take effect immediately after Christmas, instead of the more usual starting time of September. Perhaps one of the teachers had left unexpectedly, leaving them short-staffed. Filling a position partway through the academic year was probably difficult for a school, so maybe I stood a chance. I applied and was invited to attend for interview at the end of October, which was during our half-term holiday week.

Liz's due-date was Thursday, October 27th, which just happened to coincide with the day of my interview. And, right on cue, her contractions began just before six that morning. We were living with my father at the time, and the three of us sat at the breakfast table discussing whether I should keep my morning appointment or phone the school and cancel. It was decided that I would attend the interview while Dad took Liz to the hospital.

Westminster City School, located a few hundred yards from Buckingham Palace, with Westminster Abbey and the Houses of Parliament a half-mile away, is in one of the most prestigious parts of London. While imposing on the outside, the school was even more impressive from the inside, with its wood paneling and oil paintings of former headmasters. For a job applicant in his early twenties who had never gone to grammar school, it was a little intimidating.

Most of the interview was conducted by the head of the science department, John Hunt, a short, slightly tubby man in his fifties, with dark-rimmed glasses, a cheerful smile, and a propensity to guffaw aloud. We took an instant liking to one another, and he sounded well pleased with everything I told him about myself. The interview with the headmaster, a tall and impressive man named Allder, was essentially a formality, and I was offered the position. I immediately accepted. Adjourning to the staff room for coffee with John Hunt and his colleagues from the interview, I asked if I could use the telephone, explaining how my wife had gone into labor that morning. Their reaction was a mixture of amazement and awe—how could anyone be *that* calm and collected in an interview when his wife was having their first child?

When I broke the news of my new job to the head of science the following week, he was genuinely pleased for me. Knowing I still had to inform the headmaster, he said he had an appointment with him immediately following mine and asked if he could tag along, just to see his reaction. I was more than happy to oblige.

Announcing my December departure to the supreme leader was a singularly rewarding experience, especially since he took it so badly. Apoplectic with anger, he raved and ranted like a madman. My colleague was enjoying the performance almost as much as me. Then, as if realizing what a spectacle he was making, he suddenly became calm. The reason for the unexpected change was that he had just done the arithmetic.

"You're too late to give in your notice!" he exclaimed, smiling jubilantly. "If you were going to resign, you would have to have notified me before half term was over."

"I handed in my letter of resignation to the education office last week," I replied.

As the school had been closed for half-term, I had delivered my resignation to the education office in town. Going over his head by reporting directly to head office only added insult to injury. We were then treated to an encore of anger every bit as entertaining as the first.

When I started teaching at Westminster, John Hunt confided that my former headmaster had phoned the school soon after discovering I was leaving. He wanted to warn them that I was trouble. "That's just what we need!" chortled John. "Someone with a bit of spirit!"

To help familiarize me with the new material I would be teaching, John had lent me the relevant Nuffield biology textbooks to look over during the Christmas vacation.

Aside from having to master the new material and become acquainted with my new colleagues, I also had to accommodate to life at a traditional old grammar school. Many of the teachers wore academic gowns and each day began with a formal assembly in the wood-paneled hall, with all of the teaching staff seated on stage. Participation was not a matter of choice.

Back in the sixties, students were streamed into different schools according ability—grammar, technical, and secondary modern. There was also streaming within schools. The most capable students were placed in the A-stream, the less capable ones in the B-stream. There was also an R-stream, if required, for the least capable and often most disruptive students of their year, the "R" standing for *Removed*—presumably signifying their removal from the academic stream. For example, a class designated 4R would be a small and often unruly group of fifteen-year-olds, in their fourth year at school.

Streaming is now widely rejected by educationalists on the grounds that it stigmatizes students, but I saw no evidence for this back in the sixties. And I much preferred having all the disruptive elements in one class, where they only affected each other, rather than distributing them in other classes where they could disturb everyone.

As a new teacher, I had to go through a week or two of establishing boundaries, but word soon spread that I was not to be fooled with. The first to get the message was a class that ignored my warning not to talk when I was talking. When I announced that I would see them back in the classroom at the end of school, they assumed this was for a class detention. Appearing a few minutes after four, they dutifully took their places, sitting in silence while I occupied myself at the front—sorting through papers, checking in cupboards, and generally looking busy. From my casual observations, I could see their mounting anxiety: the fleeting looks at the clock, the exchanged glances, the clearing of throats. After about thirty minutes of silence, one of the boys bravely raised his hand. "Please sir," he began, "how long are we going to be kept in detention?"

"Detention?" I queried. "There's no detention tonight. I've decided to be lenient and let you all off with a warning." I went on to explain that I would talk to them as soon as I had finished what I was doing. After busying myself for another ten minutes or so, I put them on notice of the dire consequences of any further disobedience. Reminding them how lucky they were to be let off without any punishment, I dismissed them.

My unruliest class was a group of fifth-formers, designated 5S, though I have no idea what, if anything, the S stood for. Most of these fifteen and sixteen-year olds were bigger than me, and I was perceived as fair game. Nothing much happened during the first encounter, which was primarily a period of covert jousting. However, things changed significantly after that. One of them, a willowy lad with a truculent attitude, was so disruptive that I sent him outside. A minute or so later, I joined him in the corridor. Glancing up and down to make sure there were no witnesses, I engaged in a one-sided exchange that altered his attitude. That was the end of my problems with the class, and a good camaraderie developed between them and me. I still have a caricature that one of the lads penned of me—an excellent likeness, complete with my Blue Peter tie and winkle-picker shoes.[6]

The boys of Westminster came from a broad range of backgrounds. One of the lads worked on his father's stall in one of the London street markets, another was the son of a popular television actor, while a third boy's father was a minister of the Crown. A large proportion of them went on to university, and the school enjoyed a good reputation for placing students at Oxford and at Cambridge, two of the most prestigious universities in the world. Westminster was also successful in the fierce competition for places in the country's medical schools.

Knowing what my students were capable of achieving, I used to push them quite hard. "I want results not excuses!" was an oft-heard refrain in the lab. I used to have fun with them too, but they all knew when it was time to get back to work. Some teachers liked to be popular with their students, but that was not my way. On one occasion, a lad in my home form told me about a poll that had been conducted, to see who was the most popular teacher in the school. "Do you want to know where you came in, sir?" he asked, with a smile on his face.

"I would like to think I came last."

"Nearly, sir. You came second from the bottom."

"Well, I'm quite pleased to hear that, but I'm disappointed not to be last."

Rather than drinking coffee and eating our bagged lunches in the staff room on the main floor, we science teachers used to congregate in the preparation room beside the labs, along with our lab technicians. They were the ones who prepared the supplies and equipment for our classes. We all got on very well together and used to have a lot of fun, much of which was attributable to the congenial John Hunt.

Like the Roman philosopher Seneca, two thousand years before, I found the best way to learn was to teach. And one of the most valuable lessons I learned about teaching was acquired at Westminster, through teaching science the hands-on way. The lesson served me well throughout my university teaching career. However, before sharing that evocative experience, we need to return to the present and the continuing saga of teaching science in US schools.

CHAPTER 7

TREADING FAMILIAR GROUND

The Next Generation Science Standards (NGSS), based upon the *Framework* document, are essentially a series of tables describing topics to be studied in different subject areas, along with the performance expectations for students throughout grades K–12. In addition to the 102 pages of the Standards document, there is a considerable amount of ancillary material, including thirteen appendices, for a total of 445 pages.[1]

The writing team responsible for all this material was more than twice as large as that for the *Framework* document. The forty-member group comprised eighteen educationalists, sixteen science teachers (from elementary, middle, and high schools), three engineers, two scientists, and one educational consultant.[2] The influence of educationalists is apparent from the content and writing style of much of the ancillary material, as the following examples will show—an influence that has shaped the Standards document.

In a discussion of the nature of science, which is the primary theme of Appendix H, the following statement appears:

> In K–12 classrooms, the issue is how to explain both the natural world and what constitutes the formation of adequate, evidence-based scientific explanations. To be clear this perspective complements but is distinct from students engaging in scientific and engineering practices in order to enhance their knowledge and understanding of the natural world.[3]

My reading of this esoteric passage is that it is as important for students to understand what constitutes scientific evidence—hypothesis testing and all the rest—as it is to use it. Two pages further on, beneath a section titled "The Nature of Science and NGSS," the appendix states, "Here we present the NOS [Nature of Science] Matrix. The basic understandings about the nature of science are . . ." The list of eight items that follow includes such inane points as "Scientific Investigations Use a Variety of Methods," "Science is a Way of Knowing," and "Science is a Human Endeavor."

In a later section, teachers are told about having their students make some observations, like "the moon's movements in the sky," and having them "observe patterns and propose explanations of cause-effect. Then, the students can develop a model of the system based on their proposed explanation. Next, they design an investigation to test the model."[4] How plausible is it that school students could actually do this?

When I taught an annual two-week field course on marine biology for university students, I used to explain the moon's orbit around the Earth, and its effect on tides. The undergraduates, all in their third or fourth year, were obviously unfamiliar with the underlying mechanism that caused the moon's phases. I explained this graphically, using Styrofoam spheres and the light from a projector to illustrate how the position of the moon in relation to the Earth created the various phases seen during the lunar orbit. I doubt any of the students could have deduced the cause-and-effect mechanism for themselves in the way suggested above for school students. This disconnect between the theory of learning and the reality of teaching is reflected in some of the unrealistic expectations of the Standards document, a point already made with respect to the *Framework* document. The same divide between an educationalist's expectations and a teacher's reality is reflected in the concluding paragraph of Appendix H:

> Beginning with the practices, core ideas, and crosscutting concepts, science teachers can progress to the regularities of laws, the importance of evidence, and the formulation of theories in science. With the addition of historical examples, the nature of scientific explanations assumes a human face.[5]

It came as no surprise to me that all but two of the seventeen authored references cited at the end of Appendix H were written by educationalists.[6]

The Standards document can be viewed either by topic arrangement, which I chose, or with the contents arranged by DCI (Disciplinary Core Ideas). Interspersed between the numerous tables, which are tightly crammed with small print, are overviews of performance expectations for each level. These are described in the table of contents as the "storyline" for that level. There is, for example, a "Third Grade Storyline."

Without going into details as I did for the *Framework*, a few examples will serve to show the problems I have with the NGSS. Leaving aside the inclusion of engineering, earth and space sciences, the building of models, and relating science to society—all of which I dealt with in chapters 1 and 5—one of my objections is the unrealistic expectations required of students. Students in grade 1, for example, are expected to answer questions like, "What objects are in the sky and how do they seem to move?"[7] Furthermore, "Students are able to *observe* [my emphasis], describe, and predict some patterns of the movement of objects in the sky."

Six-year-olds might be able to do this for a passing airplane or a cloud, but I suspect celestial bodies are the intent here. So, are these small children expected to rise at dawn and see how the sun seems to move across the sky, perhaps until it drops below the horizon at dusk? Do they repeat these observations for the moon during the night? Then, having made their observations, are these youngsters *really* expected to "predict some patterns of the movements?" Am I missing something here, or is this complete nonsense?

By the following year, in grade two,

> Students are able to apply their understanding of the idea that wind and water can change the shape of the land to compare design solutions to slow or prevent such change.[8]

Most adults would find it difficult enough visualizing how the elements can change coastlines, but these seven-year-olds are also expected to compare designs of coastal barriers that reduce erosion. The text continues,

> Students are able to use information and models to identify and represent the shapes and kinds of land and bodies of water in an area and where water is found on Earth.[9]

I have difficulty trying to understand this seemingly implausible task. This section then concludes by stating that students are expected to demonstrate

> proficiency in developing and using models, planning and carrying out investigations, analyzing and interpreting data, constructing explanations and designing solutions, engaging in argument from evidence, and obtaining, evaluating, and communicating information.[10]

These would be demanding expectations for university undergraduates.

On the biological front, these seven-year-olds are expected to be able to answer questions such as, "How many types of living things live in a place?" They are also "expected to compare the diversity of life in different habitats."[11] Aside from being monumental tasks even for undergraduates, this suggests that some fieldwork is required. And would this not also require some knowledge of how organisms are classified? I presume so because, by the following year, "Students are expected to develop an understanding of types of organisms that lived long ago and also about the nature of their environments."[12] I cannot see how they could do this without knowing the differences between, say, ammonites, trilobites, fishes, amphibians, dinosaurs, birds, and mammals.

Some of the things youngsters are expected to know fall within the realm of the profound. Imagine middle-school students—youngsters aged between eleven and thirteen—being able to answer the question, "How does water influence weather, circulate in the oceans, and shape Earth's surface?"[13] These youngsters are also expected to answer what I consider to be the unanswerable: "What is Earth's place in the Universe?"[14]

By high-school, students are expected to "examine the processes governing the formation, evolution, and workings of the solar system and universe," which strikes me as a demanding task. They should also be able to answer the question, "How do people reconstruct and date events in Earth's planetary history?"[15]

The standards for expectations in life sciences are equally demanding, calling for students to "use statistical models to explain the importance of variation within populations for the survival and evolution of species."[16] I could certainly not aspire to do this. Nor am I sure how to go about answering the question, "How do the structures of organisms enable life's functions?"[17]

There are also some aspects of the science in the NGSS that are of concern to me. Several questions, for example, strike me as odd, like the grade four question, "What patterns of Earth's features can be determined with the use of maps?"[18] Is this not about geography? There is also a grade five requirement that, "Students are expected to develop an understanding of patterns of daily changes in length and direction of shadows, day and night . . ."[19] Barring a bright moon, would we expect to see shadows at night, other than from lampposts and headlights?

Another questionable expectation, in middle-school physics, is that, "Students will also come to know the difference between energy and *temperature* [my emphasis], and begin to develop an understanding of the relationship between force and energy."[20]

Energy is the potential to do work and has the same units as work. Given that work is force multiplied by distance, why would students need to know the difference between energy and *temperature*? The statement would have made more sense written as: Students will also come to know the difference between energy and work, and begin to develop an understanding of the relationship between work and force. This may have been a typographical error, but I think it more likely to have arisen from the same naivety of science that dogged the *Framework* document.

Another odd question, for grade four students, is "How do internal and external structures support the survival, growth, behavior, and reproduction of plants and animals?"[21] Compartmentalizing information about organisms like this seems odd to me. I wonder whether this was written by the same person who wanted to distinguish between internal and external motion in the body, as written in the *Framework* document (see chapter 5). It is difficult imagining a person well versed in science writing this way.

Given the origin of the NGSS, it is not surprising that so much attention is paid to theorizing over science, with inordinate references to models and model making and precious little importance attaching to conducting experiments. Out of interest, I counted up how many times the word *experiment* appears in the document: fourteen times in the entire 102 pages.[22] The word *model*, in contrast, appears multiple times—I stopped counting when I reached ninety-four, on page 35. The failure of this vision of teaching science is epitomized by the way that photosynthesis is dealt with. As we will see in the next chapter, photosynthesis, a process of fundamental importance to the mutual dependence of plants and animals, is readily investigated by simple experiments. Remarkably, however, the NGSS is incapable of seeing beyond expecting students to: "Use a *model* [my emphasis] to illustrate how photosynthesis transforms light energy into stored chemical energy,"[23] or to "apply mathematical concepts to develop evidence to support explanations of the interactions of photosynthesis and cellular respiration and develop models to communicate these explanations."[24]

It is this model-making, theorizing, hypothesizing approach to science that gives me the greatest problem with the NGSS. Science is a practical subject, but what we have instead is an educationalists version of reality. Whatever became of curiosity and discovering through *doing*?

I enjoy teaching, and the only thing I really miss after taking early retirement is not having students. Teaching children is especially rewarding, encouraging and directing their natural inquisitiveness. I discovered this many years ago, while teaching science in school after receiving my bachelor's degree. However, if I were graduating today, and living in the United States, I am not sure I could face complying with the NGSS to teach in school.

Imagine trying to come to grips with it all—the implausible expectations for students, the nonsensical linking and compartmentalizing of material, page upon page of tables, crammed with text and littered with codes. And, as if that were not enough, you have to teach biology, chemistry, physics, geology, and astronomy, along with engineering and technology, and link them all to society. If there were any remaining hopes of soldiering on, these would be removed by reading through Appendix K.

This appendix is titled "Model Course Mapping in Middle and High School for the Next Generation Science Standards." With its flowcharts, tables of acronyms, and directives galore, it is a singularly discouraging document. For me, figure 3 says it all.[25]

Appendix K's figure 3 is a flowchart linking thirty-nine boxes bearing such labels as "LS2.D—Social Interaction and Group Behavior," "PS3.D—Energy in Processes and Everyday Life," and "ESS1.A—The Universe and its Stars." To make everything clear, presumably, the thirty-nine boxes are linked together by no less than sixty-five arrows—I counted them.

My sympathies lie with the dedicated teachers in the United States and Canada who strive to enthuse their students with science, in spite of all the guidelines and directives they receive. I have no sympathies for those who, with little or no understanding of science themselves, formulate policy on how the subject is to be taught. They should try teaching in the classroom themselves and see how they cope with it all. Many educationalists have probably done that very thing, earlier in their careers. This brings to mind a saying, popular among UK teachers during the sixties: If you can't do it, teach it, and if you can't teach it, inspect it. Translating for this side of the Atlantic: If you cannot get a job in your chosen field after leaving university, become a schoolteacher and, if you are useless at teaching, become an administrator, or maybe an educational consultant. A brief survey of the standards documents of a few education authorities in the United States reveals some of the problems that science teachers have to face in the pursuit of their profession.

The Michigan Department of Education developed the *Michigan High School Science Content Expectations* to "establish what every student is expected to know and be able to do by the end of high school . . . in Earth Science, Biology, Physics, and Chemistry."[26] With its obscure phrases, complex coding, and unfathomable writing, the first part of this document is challenging reading. Consider, for example, the following under the heading of "Performance Expectations":

> Performance expectations are written with particular verbs indicating the desired performance expected of the student. The action verbs associated with each practice are contextualized to generate performance expectations. For example, when the "conduct scientific investigations" is crossed with a states-of-matter content statement, this can generate a performance expectation that employs a different action verb, "heats as a way to evaporate liquids."[27]

The other entry under the same heading is equally incomprehensible:

> Performance expectations are derived from the intersection of content statements and practices—if the content statements from the Earth Sciences, Biology, Physics, and Chemistry are the columns of a table and the practices (Identifying Science Principles, Using Science Principles, Using Scientific Inquiry, Reflection and Social Implications) are the rows, the cells of the table are inhabited by performance expectations.[28]

On the same page, under the heading "Content Statements," are elaborations of the four different kinds of these statements, each complete with the relevant coding. The fourth one reads,

> **Recommended science content** that is desirable as preparation for more advanced study in the discipline, but is not required for credit. Content and expectations labeled as recommended represent extensions of the core. Recommended content statement codes include an "**r**" and an "**x**"; expectations include an "**r**" and a lower case letter (e.g., **P4.r9x Nature of Light**; **P4.r9a**).[29]

This would have challenged the Enigma code-breakers of Britain's Bletchley Park, during World War II.

If there was any doubt that educationalists were behind Michigan's Science Content Expectations, it would be removed by two confusing flowcharts that appear near the beginning of the document. The first depicts a table, subdivided into boxes, with each box connecting to the one above by an arrow.[30] Each of the boxes in one of the rows is labeled "Essential Knowledge and Skills." The corresponding boxes in the row above are each labeled "Core Knowledge and Skills." Presumably, this means something to someone. The second flowchart occupies most of the following page. Crammed with small print, it has arrows going this way and that, linking boxes with names like "Models for District Alignment / Mapping," and "K–8 Educational Experience."[31] The largest of the boxes, occupying the entire top half of the chart, is labeled "Practices of Science Literacy." This is the subject of the next section of the document, which I will attempt to explain.

Four "key practices" are described in this section, namely, "Identifying," "Using," "Inquiry," and "Reflection and Social Implications."[32] The triangular diagram used to illustrate these practices (Figure 1 in the text), is usually referred to as a *pyramid of learning*. This has been used—overused according to many—in everything from educational texts to administrative meetings.[33] This particular pyramid is divided into three horizontal segments. Without going into details, the base of the pyramid is concerned with observations and data, the middle with graphs and laws, and the top with models and theories. An upward pointing arrow lies along the left side of the triangle, a downward one along the right. The four key practices are defined as follows:

Identifying performances generally have to do with stating models, theories, and patterns inside the triangle in Figure 1.

Using performances generally have to do with the downward arrow in Figure 1—using scientific models and patterns to describe specific observations.

Inquiry performances generally have to do with the upward arrow in Figure 1—finding and explaining patterns in data.

Reflection and Social Implications performances generally have to do with the figure as a whole (reflecting) or the downward arrow (technology as the application of models and theories to practical problems).[34]

This incomprehensible nonsense bears the hallmark of educationalists.

Aside from being unable to make much sense of the pages that follow, I have issues with some aspects of the science. In a segment that begins, "State or recognize correct science principles" is included the principle that "the atmosphere is a mixture of nitrogen, oxygen, and trace gases that include water vapor."[35] No scientist would consider this a scientific *principle*; it is merely a description of the composition of air.

Most of the Michigan document deals with the content and expectations in the different branches of science, of which I read the segment on biology. Things seemed quite straightforward here, and I had few criticisms. I imagine this had been written by the science teachers in the Science Work Group. The entire document should have been left in their hands, eliminating the confusing contributions of the educationalists.

The Oregon Department of Education produced its *Science Teaching and Learning to Standards* document "to provide resources and advice for teachers, professional development personnel, and teacher educators to help them implement state standards in Oregon's schools."[36] Reading through the document certainly conveyed the impression that helping teachers was the primary objective, and there were some interesting practical contributions from teachers that would prove useful to others in their profession.

One of the points made during the introduction to standards-based education is the need to change how and what is taught in science. To this end, a table is provided, showing which aspects of teaching required less emphasis and which required more.[37] I was pleased to see a shift away from lecturing and demonstrating and toward students actively engaged in investigations themselves. However, the shift toward "responding to individual student's interests" and "sharing responsibility for learning with students"[38] is counterproductive. In this educationalist-inspired philosophy, teachers are encouraged to tailor their teaching to the particular interests of their students, thereby giving them some responsibility for lesson content. The folly of this strategy was graphically demonstrated by my experience at my grandchildren's school with youngsters telling me what *they* knew about dinosaurs (see chapter 5). The same teaching-by-consensus notion is repeated a few pages later: "Fostering students' understanding requires paying attention to the ideas they already have coming in to the class—both ideas that are incorrect and ideas that can serve as a foundation for subsequent learning."[39]

One section that puzzled me deals with the reading of science texts, and how teachers should help students to "approach a reading assignment as you would an inquiry investigation. Create an environment of discovery; one that promotes risk-taking."[40]

I fail to see how risk-taking comes into reading a book, but the theme continues:

> Make a connection between the text and your life, your prior knowledge of the world, or another text. Think of both big and little ways that you have experienced feelings, thoughts, actions, concepts and situations that are taking place in the text.[41]

How could *any* of this be relevant to a textbook on school science? But there is more: "Develop questions about the text, the author and yourself related to your reading . . ."[42] And, if it had never crossed your mind to do so, you could, "Slow down when reading confusing passages of text." And even, "Read back over the portion of the text that you are having trouble understanding."

If, like me, you wonder who comes up with such absurdities and what could possibly come next, this resource book for science teachers introduces readers to the Learning Cycle: "The learning cycle is an instructional model that can be used to facilitate scientific inquiry."[43] The objective is "to help students develop a concept, deepen their understanding of the concept, and apply the concept to new situations (Beisenherz & Dantonia, 1996)."

The Learning Cycle is illustrated by a flowchart—that almost obligatory part of every educationalist's presentation. In checking the biographies of these two authorities on education, I was intrigued to discover that, while Dr. Paul Beisenherz is a retired professor of education at the University of New Orleans, Dr. Marylou Dantonio, who also holds a doctorate in education, is a reflexologist, Reiki master, and energy intuit who "integrates reconnection energy processes in her healing sessions."[44] That a practitioner of pseudoscience should be one of the cited authorities in a resource book for science teachers underscores the problem I have with educationalists having anything to do with the teaching of science.

The Learning Cycle leads into the Learning Cycle Planner and then into the concept of *Design Space*. Here is yet another educationalist inspired "tool" for the teacher to tackle.[45] The Design Space is illustrated for readers by a diagram showing three overlapping circles, labeled "Experience," "Scientific Inquiry Skills," and "Science Knowledge." The Design Space is labeled as the shaded area where the triad of circles overlap. By the time teachers have finished reading about the Design Space, they are not even halfway through the document. What then follows are pages upon pages of worksheets, charts, assessments, decision points, threads, scoring guides, assessment tools, and more. And that is not all, because there is then a second teacher's resource document to read, all about engineering.

Oregon is seriously committed to integrating engineering and science in the classroom and, to this end, teachers are provided with a comprehensive teaching guide, the *Teacher's Guide to Using Engineering Design in Science Teaching and Learning*. This guide is available in middle-school and high-school editions. The rational for creating the guide is given in the opening statement: "Educators in many states are discovering that engineering is a great way to support, improve and enhance the teaching of science."[46] Regardless of my skepticism about anything educators have to say, I have already given reasons why engineering, as a separate entity, should not be included in school science (see chapter 5). I extend my reasons to include *engineering design*, the methodology engineers use. Engineering design is referred to extensively in the NGSS, as it is in the *Framework*.

Several pages after the opening statement, Oregon teachers are informed that

> According to Cary Sneider, a leading science educator and one of the writers of the Next Generation Science Standards, understanding engineering is essential for all citizens, workers, and consumers in a modern democracy.[47]

While I believe this overstates the case, engineering is an important field of study and a good case could be made for teaching it at as a separate school subject.

After introducing and justifying engineering design in the classroom, attention is drawn to how to teach the subject to students. Among the seven steps listed for teachers to follow is the remarkable directive to "start out by separating the boys and girls," as this "may overcome culturally differentiated experiences."[48] What is going on here?—Am I missing some new incarnation of political correctness? The guide continues,

> In the early stages of developing spatial analysis skills and learning to build, girls may do much better if you start them out in an all-girl environment. After the girls and boys have been successful and built some self-confidence, having both genders on each team will help both genders. An even mix is best, even for professionals.[49]

If my understanding is correct, girls are separated from boys in the early stages of engineering design. Then, after they have built up some self-confidence, the genders are mixed. The teacher's guide I have been quoting from is the middle-school edition and I assumed that, by the time students moved on to high school, the genders would already be mixed. However, the same directive to separate the sexes is repeated in the high-school edition. This makes little sense, but so much of Oregon's science education program puzzles me.

Washington State, Oregon's neighbor to the north, pays scant attention to engineering design in its *K-12 Science Learning Standards*. This came as a surprise to me because Dr. Cary Sneider, a staunch advocate of the importance of engineering, is acknowledged for his consulting services. His name also appears on the cover page as one of the two educationalists who prepared the document, and he was one of the team leaders in the writing of the NGSS. In the overview to Washington's standards document, readers learn that it

> strengthens the foundations of the previous document and incorporates the latest findings of educational research. The earlier document was based on three Essential Academic Learning Requirements (EALRs). In the new standards, EALRs 1, 2, and 3 describe crosscutting *concepts* [my emphasis] and abilities that characterize the nature and practice of science and technology.[50]

The references to educational research, the use of acronyms, and the inclusion of a flow-chart portend the influence of educationalists, and this is confirmed by the incomprehensible material that follows.

EALRs 1–3 are identified as *Systems*, *Inquiry*, and *Applications*, respectively. Given that the word *concept* is defined in the glossary as a "universal idea,"[51] how could these three items possibly be included here as concepts? The next question that comes to mind is how these terms are being used in the document.

Contrary to any dictionary definition I could find, the glossary defines *system* as: "An assemblage of interrelated parts or conditions through which matter, energy and information flow."[52] Thus defined, it could apply to anything from a beehive to a bank machine. Similarly, *inquiry* differs from any dictionary definition: "The diverse ways in which people study the natural world and propose explanations based on evidence derived from their work."[53] *Application* is not included in the glossary, but *Apply* is defined as: "The skill of selecting and using information in new situations or problems."[54] Again, this is a departure from any dictionary definition.

The following descriptions of EALRs 1-3 only reaffirm the absurdity of referring to them as *concepts*:

> **EALR 1 Systems** thinking makes it possible to analyze and understand complex phenomena. Systems concepts begin with the idea of the part-to-whole relationship in the earliest grades, adding the ideas of systems analysis in middle school and emergent properties, unanticipated consequences, and feedback loops in high school.
>
> **EALR 2 Inquiry** is the bedrock of science and refers to the activities of students in which they develop knowledge and understanding of scientific ideas, as well as an understanding of how the natural world works. . . . Inquiry includes the idea that an *investigation* refers to a variety of methods . . . including: systematic observations, field studies, models and simulations, open-ended explorations, and controlled experiments.
>
> **EALR 3 Application** includes the ability to use the process of technological design to solve real-world problems, to understand the relationship between science and technology and their influence on society, and . . . contribute to the prosperity of their community, state, and nation.[55]

These three "concepts," *systems*, *inquiry*, and *application*, are subsequently referred to as "Big Ideas," for the following reason:

> The strategy of using Big Ideas to organize science standards arose in response to research showing that U.S. students lagged behind students

> in many other countries, at least in part because school curricula include far too many topics . . .
>
> A solution to this problem that has gained support from science education researchers in recent years is to organize science standards by a small number of "Big Ideas," which are essential for all people in modern society to understand.[56]

No references to sources are given in this document, but I assume that this is the same "big ideas" encountered earlier (chapter 1), attributed to Wiggins and McTighe.

The last EALR is described as follows:

> **EALR 4 The Domains of Science** focus on nine Big Ideas in the domains of Physical Science, Life Science, and Earth and Space Science that all students should fully understand before they graduate from high school so that they can participate and prosper as citizens in modern society.[57]

The nine big ideas of EALR 4, together with the three for EALRs 1–3, makes for a round dozen: "Each 'Big Idea' is a single important concept that begins in the early grades, and builds toward an adult-level understanding."[58] The twelve big ideas are listed and elaborated upon in the document's Appendix A.[59] They are: 1. Systems; 2. Inquiry; 3. Applications; 4. Force and Motion; 5. Matter: Properties and Change; 6. Energy: Transfer, Transformation, and Conservation; 7. Earth in the Universe; 8. Earth Systems, Structure, and Processes; 9. Earth History; 10. Structure and Functions of Living Organisms; 11. Ecosystems; and 12. Biological Evolution.

With the exception of biological evolution, none of these can be considered as concepts. Furthermore, most of them comprise multiple parts. For example, the eighth big idea, Earth in the Universe, encompasses everything from the phases of the moon and seasons on Earth to distant galaxies and the big bang theory. The inclusion of so many different subjects is contrary to the rationale given for using big ideas, which was that "school curricula include far too many topics."

Referring to other science standards across the United States, it is stated that

> Although most state and national standards include the domains of science and scientific inquiry, and the application of science and technology to society, Washington is unique in emphasizing *systems*.[60]

The reason given for selecting *systems* includes helping students with "some of the challenges they encounter in everyday life as citizens, workers, and consumers."[61]

No doubt many educationalists would find much to praise in Washington State's *Science Learning Standards*. However, all I see is the usual esoteric writing about science and how it should be taught that does more to hinder than help. This is abundantly clear in the Science Standards section. The purpose of the standards is to set out what students are expected to know at each level in the educational system. The point is made, however, that the standards should not be used to limit science programs and that, "Young children should have many experiences to spark and nurture their interests in science."[62] With so much attention paid to hypotheses, conceptual models, systems analysis, and the like, youngsters would find little to arouse their interests. Consider, for example, the following segment, on the Big Idea of Inquiry. Here, nine and ten-year-olds are expected to: "Create a simple *model* to represent an event, *system*, or process. Use the *model* to learn something about the event, *system*, or process. *Explain how* the *model* is similar to and different from the thing being modeled."[63] (The significance of italics, here and elsewhere, is not apparent to me.)

Students are also expected to: "Conduct or *critique* an *experiment*, noting when the *experiment* might not be fair because things that might change the outcome are not kept the same."[64] Between the ages of eleven and thirteen, they are expected to, "*Generate* a scientific *conclusion* from an *investigation* using inferential logic" and to "*describe* the differences between an objective summary of the findings and an *inference* made from the findings."[65] Students in their teens are told that, "The essence of scientific *investigation* involves the development of a *theory* or conceptual *model* that can *generate* testable predictions."[66] To this end, presumably, students are expected to, "Formulate one or more *hypotheses* based on a *model* or *theory* of a causal *relationship*. Demonstrate creativity and critical thinking to formulate and *evaluate* the *hypotheses*."

With all this attention to the methodologies of science, do students ever get to do any hands-on investigations themselves? Not a lot, it would seem.

However, while studying the Big Idea of Force and Motion, nine and ten year-olds do get to "Use a spring scale to measure the *weights* of several objects accurately."[67] This is so they can explain that the weight of an object is a measure of the force of gravity acting upon it. They also get to "Measure the time it takes two objects to travel the same distance and determine which is fastest.*"[68] The asterisked item reads, "Estimate and determine elapsed time, using a *calendar* [my emphasis], a digital clock, and an analog clock." I cannot imagine the significance of the calendar, unless it is to measure the apparent duration of this unbelievably boring lesson.

Students studying Force and Motion a couple of years later, get to, "Illustrate the *motion* of an object using a graph, or *infer* the *motion* of an object from a graph of the object's position vs. time or *speed* vs. time."[69] They also get to conduct an experiment where they measure the distance traveled in a given time by an object, like a battery-powered car, so they can measure its average speed. It is difficult to imagine any of these activities sparking an interest in science.

If I were teaching youngsters about force and motion, I would get them to launch toy cars across the floor with an improvised rubber band device.[70] By using a spring balance to draw back the cord, the youngsters could get a measure of the strain energy stored in the device prior to each launching. This would allow them to compare the distances traveled for different amounts of energy. It would also get them thinking about the relationships between force, mass, work, and energy. This, of course, requires an understanding of the term *energy*.

In the Appendix entry that describes the Big Idea of Energy, it is said that: "Although it is difficult to define, the concept of energy is very useful in virtually all fields of science and engineering."[71] To say that *energy* is difficult to define is sheer nonsense. Energy is simply defined as the potential to do work, and it has the same units as work; work is force times distance.[72] Energy is incorrectly defined in the glossary as "The amount of work that can be done by a force."[73]

While on the subject of getting the science right and using appropriate terminology, I have a few more quibbles. In a segment on the Big Idea of Matter, it is said that "*Air* is a *gas*."[74] To be accurate, air is a mixture of gases, mostly nitrogen and oxygen. It is also said that students are expected to "*Explain* that water is still the same substance when it is frozen as ice or *evaporated* and becomes a gas." Clearly, water is not the same substance as ice, which is why you cannot skate on a pond that is not frozen. The error here is that the word *compound* should have been used, not *substance.*

Washington State is obviously proud for being unique in emphasizing *systems* in its Science Learning Standards. From my perspective, however, this particular Essential Academic Learning Requirement is the most ridiculous of them all. Imagine being a teacher, standing in front of a class of nine or ten year olds and explaining that "Systems contain subsystems. A system can do things that none of its subsystems can do by themselves."[75] Now that you have riveted their attention, you can elaborate: "One defective part can cause a *subsystem* to malfunction, which in turn will affect the *system* as a whole." You could illustrate this with a picture or an airplane—they have *lots* of systems. If the engines fail, or the landing gear, or those wiggly flaps on the wings, then the whole airplane is affected. There is even more to tell: "Systems have *inputs* and *outputs*. Changes in *inputs* may change the *outputs* of a *system*."

By the time you have finished, your students should be able to explain about inputs and outputs and give an example. For instance, "when making cookies, *inputs* include sugar, flour, and chocolate chips; *outputs* are finished cookies."[76] They should also be able to describe "the *effect* on a *system* if its *input* is changed (e.g., if sugar is left out, the cookies will not taste very good)." If this does not spark their interest in becoming a scientist, maybe some will become bakers!

Students in middle school (eleven to thirteen years old) learn that systems can be open or closed: "In an *open system, matter* flows into and out of the *system*. In a *closed system*, *energy* may flow into or out of the *system*, but *matter* stays within the *system*."[77] This information is followed by the revelation that: "If the *input* of *matter* or *energy* is the same as the *output*, then the amount of *matter* or *energy* in the *system* won't change; but if the *input* is more or less than the *output*, then the amount of *matter* or *energy* in the *system* will change." I can imagine the gasps of wonder at this! And the students get to conduct a hands-on experiment: "Measure the flow of *matter* into and out of an *open system* and *predict* how the *system* is likely to change (e.g., a bottle of water with a hole in the bottom . . .)." If anyone had set out with the specific intent of turning youngsters off science, they could hardly have done better that introducing the absurdity of *systems* into the classroom. That one of the two educationalists who prepared Washington's standards document, Dr. Cary Sneider, was one of the team leaders in the writing of the NGSS should not pass unnoticed.[78]

The damage done to school science by the micromanaged meddling of educationalists is epitomized for me by the following student expectation: "Given a *system*, identify *subsystems* and a larger encompassing *system* (e.g., the heart is a *system* made up of tissues and cells, and is part of the larger circulatory *system*)."[79] Here, it is more important to think of the heart as an example of a system of subsystems, set within a larger system, than as a living functional organ. An organ that pumps blood around the body, through a series of chambers and valves, thirty million times or more a year, from before we were born until we take our last breath.

I have a mental picture of a class at the school where I taught, back in the sixties, dissecting fresh cow's hearts, probing and exploring and discovering. Do students do such things anymore? I searched Washington's *K–12 Science Standards* for *dissection* and found nothing. I tried again for *skeleton*, with the same result. Then I tried for *anatomy.* I found two matches, both for the *Anatomy of a Standard.*

I repeated the same exercise for the NGSS document with similar results: nothing for *dissection* or for *skeleton*, and two entries for *anatomy*. One of these entries made reference to the fact that embryos of different species show greater similarities in their anatomy than do fully formed individuals. The other entry reads: "Analyze displays of *pictorial data* [my emphasis] to compare patterns of similarities in the embryological development across multiple species."[80] Not only do students have limited opportunities for conducting experiments themselves, but they do not get to see *real* specimens either.

We are told that "*A Framework for K-12 Science Education* casts a bold vision . . . and the resulting Next Generation Science Standards (NGSS) have taken a huge leap toward putting this vision into practice."[81] We also learn that "One fundamental goal for K–12 science education is a scientifically literate person."[82] How can this be so when those who have had most influence on the *Framework* and NGSS, the educationalists, are not scientifically literate themselves?

It troubles me to think that the youngsters I taught, over half a century ago, had a better understanding of science than children are receiving in our schools today.

CHAPTER 8
THE NUFFIELD PROJECT

During the late fifties and early sixties, science teachers in the UK established a working group to devise new and better ways of teaching science. What they wanted was to break away from the traditional approach, where teachers did the talking and demonstrating of practicals while pupils passively listened and learned. The Nuffield Foundation, a charitable trust, had also been investigating the problem and agreed to fund and coordinate the process. This resulted in the Nuffield Science Teaching Project, which began in 1962. It was decided that the best way to engage students in science was to have them explore and discover things for themselves, learning through hands-on practical experience.

There was no standard Nuffield approach to teaching science, and three separate teams of teachers—for biology, chemistry, and physics—worked largely independently. Professional scientists were available to help the teams, and a large number of schools across the country took part in trials and in adopting the new Nuffield curriculum. Each of the three groups produced a set of textbooks for students, along with a companion series of teachers' guides, which I found particularly useful.

The first Nuffield biology class I taught (equivalent to grade 7) began with a unit on bacteria and disease. This introduced the students, and me, to Louis Pasteur (1822–1895) and his classic experiment showing that bacteria existed everywhere in the world, floating in the air. Back in Pasteur's time, there were two competing ideas to explain what caused things to spoil, like milk souring, meat rotting, and wine turning to vinegar. According to some, this happened spontaneously when harmful microbes suddenly appeared where there had been no microbes before. These minute organisms had already been seen by studying droplets of water beneath the primitive microscopes then available.[1] This process, called *spontaneous generation*, was also thought to be the cause of infectious diseases, like cholera and typhoid, which were rampant in the overcrowded and unsanitary towns of the day. Other people, Pasteur included, believed that these harmful beings—now referred to as bacteria—floated in the air, along with dust particles. This idea was called the *germ theory*.

Figure 8.1: Louis Pasteur (1822–1895).

Pasteur conducted a simple experiment to test which idea was correct, using a pair of round glass flasks with long necks. Adding some beef broth to the first flask, he heated the neck with a Bunsen flame, melting the glass and drawing it out into a long thin S-shaped bend, like a swan's neck. After boiling up the broth to kill any microbes, he set it aside to cool down. As the air inside the flask cooled and contracted, air was drawn inside, passing through the narrow bends slowly enough for the dust particles and microbes to settle. Air was therefore free to enter flask and reach the broth, but all the particles were trapped inside the neck. He repeated the same procedure with the second flask but, after boiling up the broth, he snapped off the neck so that it was straight. Air, along with dust particles and microbes, could now freely enter the flask and reach the broth. After several weeks, the broth in the second flask had turned bad, but the broth in the swan-neck flask looked the same as it had at the beginning of the experiment.

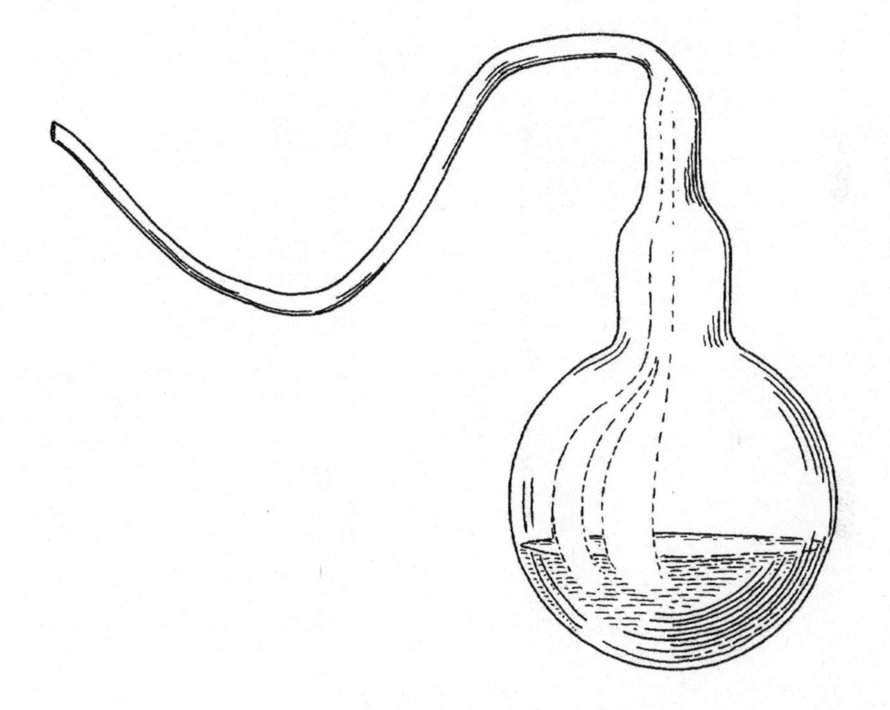

Figure 8.2: Pasteur's swan-necked flask apparatus.

After telling my class of twelve-year-old boys about Pasteur and the germ theory, they broke up into small groups to repeat his experiment themselves. Instead of using large round flasks with long slender necks, they used a pair of large test tubes fitted with two different kinds of stoppers. One type comprised a rubber plug fitted with a glass tube bent into an S-shaped curve. The other was a plug with a straight length of glass tube. After boiling each of the test tubes containing broth over a Bunsen burner, each plug was pushed tightly into place and the tubes transferred to a rack. Having labeled the tubes with the time, date, and the name of their group, the racks were left where they could remain undisturbed for several weeks. A class discussion followed on what was likely to happen, and why. Similar discussions continued during the weeks that followed as each group monitored the changes taking place.

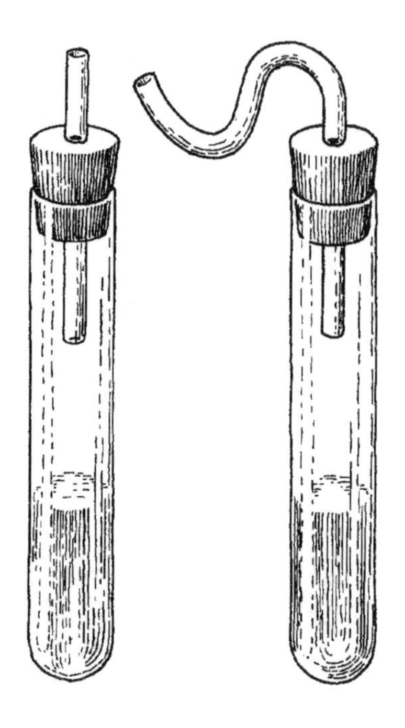

Figure 8.3: A classroom substitute for Pasteur's original experiment, using test tubes fitted with rubber plugs, with a straight and with a curved glass tube.

Back then, we had lab technicians to make up things like nutrient broth, and to create equipment like S-shaped glass tubes, but today's teachers are largely unassisted. But there are ways of devising experiments that require minimal preparation. Pasteur's experiment can be repeated using nothing more elaborate than a pair of clear wine bottles and access to a microwave oven. After adding about 60 ml (4 tablespoons) of milk to each bottle, they are laid, side by side, on a plate, covered with crumpled paper towel to prevent them rolling around. Setting the microwave for two minutes at full power boils the milk, killing the bacteria and sterilizing the bottles with the steam. While allowing the bottles to cool down inside the closed oven for ten minutes, a plug is made for one of them by folding and rolling together a pair of paper tissues. On opening the door, the plug is quickly pushed into the neck of one of the bottles, like a cork. Both bottles are then stood upright and left somewhere where they can remain undisturbed for several weeks. The paper plug allows air to enter but traps the dust and other particles.

Within a week or so, the milk in the open bottle starts clotting and begins to smell, but the other milk looks unchanged. After a month, I found that the milk in the open bottle smelled awful and was brown and clotted. However, the other milk looked essentially unchanged, even after two months.[2]

Figure 8.4: A simple repetition of Pasteur's experiment, using a pair of bottles. Even after two months, the milk in the plugged bottle looks unchanged, whereas that in the open bottle looks terrible.

Teaching Nuffield biology during those first few weeks was challenging on all fronts. Not only was there so much unfamiliar material, but I also had to adapt to the new approach to classroom science. I enjoyed talking as much as teaching, so I always had to be mindful of the time when standing in front of a class. And there I was at Westminster, talking for mere minutes before leaving my pupils to their own devices, with minimal help from me. Was Nuffield biology the way to go? Were these boys really going to understand about bacteriology? Would it not be better for me to lecture them on Pasteur and the germ theory, and about Fleming and penicillin, rather than having them spend most of their time experimenting?

Empirical by nature, I have to see things and test things before being convinced. And what I was seeing was youngsters catching on fast, knowing exactly what they were doing and what it was all about. They were not memorizing and recalling facts from the textbook but learning and understanding from first-hand experiences. They were also *motivated.* Teaching at Westminster taught me a valuable lesson: never underestimate what youngsters can achieve if properly taught and motivated.

My students learned how Pasteur's research had such an enormous and positive influence on France's agricultural industry. This was especially true for her wineries, which were in serious trouble because so much wine was going bad during the ageing process. Pasteur resolved the problems by instructing vintners to heat their wine to about 55°C (131°F) for a short time. This killed most of the bacteria without affecting the wine's flavor, or its appearance. The same process was also used for milk and other foodstuffs, greatly extending their shelf life. This was particularly important during Pasteur's time because there were no fridges to keep things fresh. The process, named *pasteurization* in his honor, is still used today.

Having set up their experiments to show that microorganisms were freely floating in the air, the students began a series of investigations to inoculate Petri dishes with bacteria. These inoculations took place under a wide range of experimental conditions. Aside from being really engaged in what they were doing, the students learned a great deal of practical information from the experiments. And so did their teacher.

Students were provided with sterile Petri dishes, the bottoms of which had been coated with a thin layer of nutrient agar, a jelly that is solid at room temperature. After exposing the dishes to bacteria, the lids were replaced and the dishes transferred to an incubator, maintained at about 37°C (98°F) to encourage bacterial growth. As the bacteria multiplied on the agar they formed colonies, which could be seen as thin blobs and streaks that spread across the bottom of the dish. These colonies appeared within one or two days of incubation.

Working in small groups, the students conducted a number of experiments to find out where most bacteria could be found. The simplest experiment involved exposing Petri dishes in different locations, leaving off the lids for a set period of twenty minutes. The locations could be on top of a cupboard, on the floor beside the cupboard, in a drafty area of the lab, in a part of the lab not exposed to drafts, and outdoors. After exposing the dishes, the lids were replaced and appropriately labeled with the location, date, and the name of the group. In addition to exposing the Petri dishes to bacteria, one dish was left unopened to act as the control, and duly labeled as such.

The importance of having controls for scientific experiments was discussed in class by giving some examples. If it was claimed, for example, that spraying seedlings with a special additive made them grow faster, this could only be verified by a controlled experiment. Several batches of seeds would be grown in similar trays and placed in similar locations. Half of the seedling would be sprayed with plain water—the controls—the other half with water containing the additive. Only if there were significant increases in the growth of the specially treated ones could the additive be said to be effective. By having an unopened Petri dish for comparison in the experiments, any differences in bacterial growth in the other dishes could be attributed to what they had been exposed to while the lids were removed.

Other experiments included having one person in the group open the lid of one of the dishes just enough to insert a finger. After gently touching the agar to leave a fingerprint, the lid was quickly closed. A labeled ring would then be drawn around the print, on the underside of the dish, to identify this as an unwashed finger. A second student would then repeat the experiment, after having first washed his hands thoroughly. Fingers could also be wiped across the bench, and elsewhere, before making a fingerprint. This, of course, required careful pre-washing of the student's hands. And care had to be taken in all these experiments not to breathe anywhere near any open Petri dishes, as this could introduce unwanted additional bacteria.

Some youngsters tried coughing onto an exposed Petri dish. There were also many other possibilities that could have been explored. These include dabbing the inside of the nostril with a cotton swab and touching this against the agar; rapidly ruffling the hair over an open dish for several seconds; and holding a text book over an open dish, spine side up, and fanning through the pages several times. There was much anticipation of what would be discovered at the end of the incubation period, and I was as intrigued as my students.

When everyone had finished examining the results of their own experiments, these were compared with those of the other groups. A class discussion would follow and one of the topics might be the importance of hand-washing. If these experiments were repeated in today's classrooms, hand-washing would likely be the number one topic.

Students explored the efficacy of penicillin by adding small paper discs, impregnated with the antibiotic, to Petri dishes that had been inoculated with bacteria. After a short incubation period they were able to see clear circular zones in the bacterial colonies around the paper discs, where the penicillin had inhibited bacterial growth.

A second method of inoculating a Petri dish is to streak samples of liquid containing bacteria across the nutrient agar, using an inoculating loop. This is simply a small loop made at the end of a wire that has been attached to a wooden handle. Prior to dipping the loop into the liquid, it is sterilized in a Bunsen flame for several seconds. Among the many liquids that can be checked for bacteria are milk, saliva, blood from meat, and water from ponds, puddles, and fish tanks.

In addition to seeing the spread of bacterial colonies across Petri dishes, the students were able to examine the organisms with a microscope. Bacteria occur in all bodies of water, but a readily available and convenient source of bacteria is water that has been shaken up with soil and allowed to settle for a few days.

How feasible is it to conduct bacteriological investigations in today's classrooms? Petri dishes are readily available from biological suppliers, along with the nutrient agar for plating them. Sterile, pre-plated Petri dishes are also available, eliminating preparation time, but these are more expensive and cannot be re-sterilized. The largest obstacle to teaching bacteriology in schools is likely to be the cost of the incubator, but this equipment can easily be improvised, at minimal cost.

I make mead, a delectable libation made from honey, which has been enjoyed for millennia. Fermentation is inordinately slow, and I keep a five-gallon carboy—a large glass jar with a narrow neck—in a homemade incubator for more than a year before the contents are golden clear for bottling. My initial source of heating for the plywood box was a 60-watt light bulb, controlled by a thermostat to maintain a temperature of about 35°C (95°F). However, when the thermostat failed, I found that a 15-watt bulb, left on continuously, worked perfectly well. Here is a simple and inexpensive way to make an incubator for introducing students to the intriguing world of bacteriology.

Some of the experiments my students conducted were things I had not done as an undergraduate, far less as a schoolboy. In learning about digestion, for example, the pupils made a working model of the gut, from a length of a particular kind of plastic tubing. This material, which is as thin as plastic wrap, is called Visking tubing in the UK, and dialysis tubing in North America. Like the gut, this membrane is semipermeable, meaning that it allows the passage of small molecules, but not of large ones.

Working in small groups, the students began by cutting off about six inches (15 cm) of the one-inch-wide (25-mm-wide) tubing. After tying off one end with thread, the tube was filled with a mixture of starch and glucose solutions. The top end was then secured with another piece of thread and the tube thoroughly rinsed off to remove any traces of the solutions. Placing the tube into a small beaker of water, this was kept at body temperature by surrounding it with tepid water inside a large beaker.

The objective of the experiment was to see whether the membrane acted as a barrier for the starch or for the sugar. To do this, we had two separate solutions that could be used to detect for the presence of sugar and for starch. One solution, called Benedict's reagent, is blue and turns red in the presence of glucose and similar sugars. The other, a solution of potassium iodide and iodine, turns dark blue or black when exposed to starch. Using a dropping pipette, the students collected small samples of water from the surface of the Visking tube every few minutes, testing these in turn for starch and for sugar. Within about twenty minutes, they found glucose in the water, but no starch. Visking tubing, like the gut, is permeable to sugar molecules, which are small, but not to the larger starch molecules. Having established that the tubing could be used as a model for the gut, they were ready to conduct an experiment on digestion.

Saliva contains an enzyme, ptyalin, which digests starch, forming sugar. Students were able to investigate the effectiveness of their own saliva by using the model gut. To do this, a student in each group collected a sample of his saliva by washing out his mouth to remove any residual sugars, and then chewing on a clean rubber band for a few minutes. (Chewing on something in the mouth stimulates the flow from the salivary glands.) Meanwhile, the rest of the group prepared another length of Visking tubing, filling it three-quarters full with starch solution. After topping this up with saliva and sealing the tube, it was shaken to mix the contents. Once rinsed clean under a tap, the tube was placed into a fresh beaker of water. Keeping the water at body temperature for half an hour, this was then tested for sugar and for starch. I was as impressed as the students when they detected sugar in the water surrounding their Visking tube models.

Sugar, the primary source of energy in the body, is broken down (oxidized) in the cells by a chemical process called cellular respiration. This process involves interactions between oxygen, water, and sugar, and results in the release of carbon dioxide, which is breathed out. Cellular respiration can be detected by the release of carbon dioxide.

This class, of thirteen-year-olds, used two different methods to see if they could detect carbon dioxide in their exhaled air. In the first experiment, a large glass gas jar was inverted over a beehive shelf in a trough of water, as if it were being used in a chemistry experiment to collect gas by the downward displacement of water. Using a length of plastic tubing threaded all the way through to the top of the jar, one student in each group placed the free end of the tube in his mouth and breathed out, filling the gas jar with exhaled air. Then, making sure not to let the water level inside the jar rise as high as the open end of the tube, he breathed in and out for as long as he could. By finishing up with an outward breath, the jar was left full of re-breathed air. While keeping the mouth of the jar beneath the water, it was covered with a glass lid to seal the contents. The jar was then removed from the water, inverted, and stood on the bench. A match was struck and, once it was burning brightly, the lid was slid off the jar and the match inserted inside. The match was immediately extinguished by the carbon dioxide. Since this gas is denser than air, it initially remains inside the jar. I could also have demonstrated this to the students by lighting one of those short candles, lowering it to the bottom of a beaker, and then extinguishing the flame by pouring the gas into the beaker from the jar, as if it were water.

This experiment can be repeated at home simply by breathing in and out of a large air-tight plastic bag. Once you have a bagful of re-breathed air, you can test this with a match. Alternatively, you can light a candle at the bottom of a large jug and extinguish the flame by squeezing out the contents of the bag, gently enough to avoiding blowing out the candle with the draft.

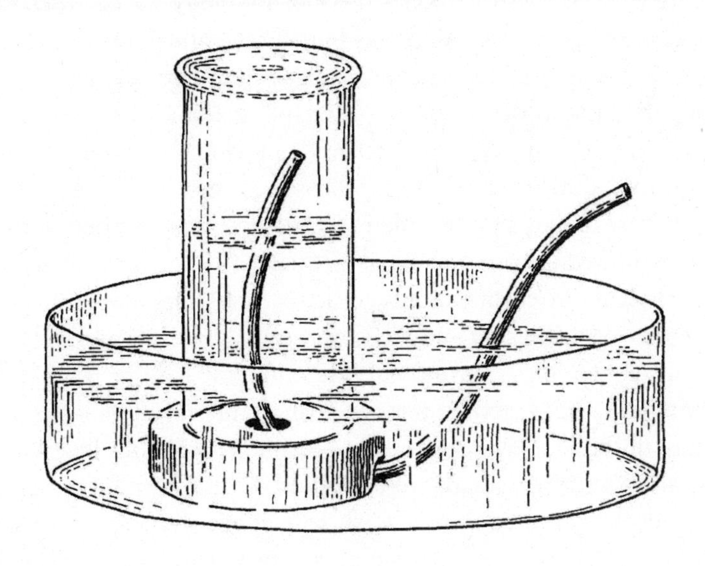

Figure 8.5: The apparatus my school students used in the 1960s to collect samples of their re-breathed air.

In the second experiment to detect carbon dioxide, the students bubbled their exhaled air through limewater (calcium hydroxide) at the bottom of a test tube, using a glass tube. Carbon dioxide reacts with limewater to form chalk (calcium carbonate), which is insoluble, causing the limewater to become cloudy with minute chalky particles.

Having shown that exhaled air contains carbon dioxide, from cellular respiration, the next question was whether this process also occurs in other living things. How could we find out? The question might have evoked some interesting class discussion, but the textbook offered a simple solution.

This experiment involved placing a small animal—such as a beetle, woodlouse, caterpillar, worm, spider, snail, or slug—into a large test tube, fitted with a wire-gauze platform placed a couple of inches (5 cm) below the top, to hold the animal. At the bottom of the test tube was about half a teaspoon (2 ml) of limewater. For comparison, some groups used bicarbonate indicator solution instead, which changes color according to the acidity. When carbon dioxide dissolves in water it forms an acid, carbonic acid—this is what gives carbonated water its tangy taste. When carbon dioxide levels are low, the solution is purple, changing to orange and then to yellow as levels rise and the solution becomes more acidic.

Once the animal had been placed inside the test tube, it was sealed with a tight fitting plug. A second tube, identical to the first except for not having an animal, was sealed tight at the same time, to act as a control. By keeping the two tubes side by side in a test-tube rack, the only difference between them was the presence of an animal.

The students in each group recorded which animal was in the tube (snail, worm, or whatever) and the time when the tube was sealed. Then, taking care not to splash the animal inside, the tubes were gently shaken, every minute, to help mix the trapped air with the solution at the bottom. When any differences were seen between the solutions in the two tubes, this was noted and the animals were released. If time permitted, this experiment would be repeated with a fresh pair of tubes and another kind of animal. The results from each group of students were written on the board so everyone could share the information. These experiments demonstrated that animals of all different kinds produced carbon dioxide, showing that animals respired. But what about plants?

Although the students had not yet studied photosynthesis, many of them probably knew that plants gave off oxygen when exposed to light. One of the easiest ways for them to verify this was to shine a bright light onto a clump of water weed in a tank or large beaker. By covering the plant beneath a glass funnel, they could collect the gas bubbles that formed, using an inverted test tube that had been filled with water. Once the gas had displaced all the water, the tube was sealed and removed from the tank. Testing the gas with a glowing splint caused the wood to burst into flame, confirming it was oxygen.

While this experiment convincingly showed that plants produce oxygen when exposed to light, it raised the question of what happens when there is little or no light.

To investigate this, each group set up a series of four large test tubes, each with some bicarbonate indicator solution at the bottom to test for carbon dioxide. Three of the tubes contained a green leaf, all of equal size. By jamming the stalk of each leaf up against the rubber plug used to seal the tube, the leaf was kept clear of the solution. The fourth test tube was left empty, as a control. One of the tubes containing a leaf was left

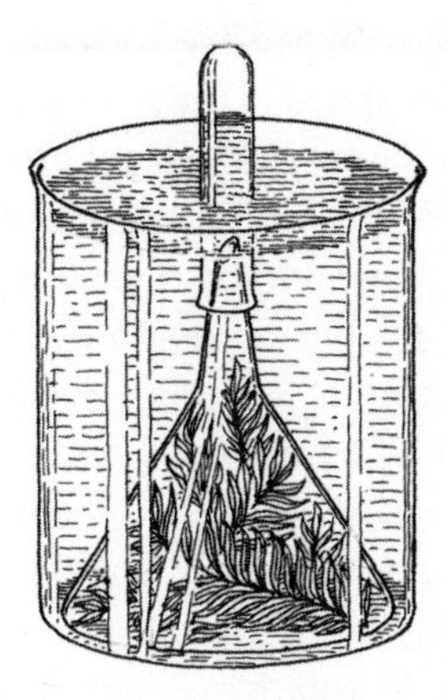

Figure 8.6: Collecting the oxygen generated by freshwater plants when exposed to light.

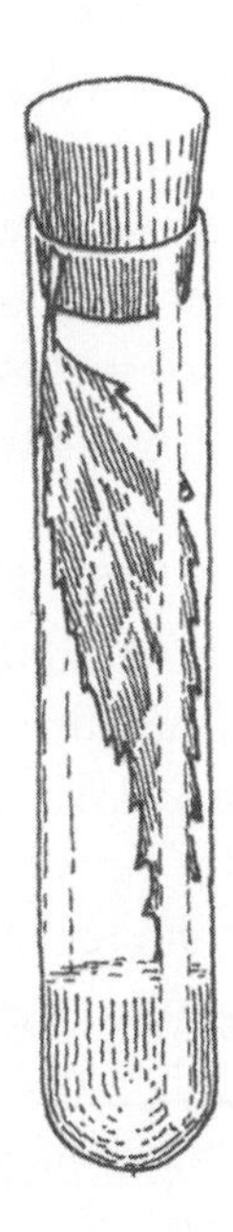

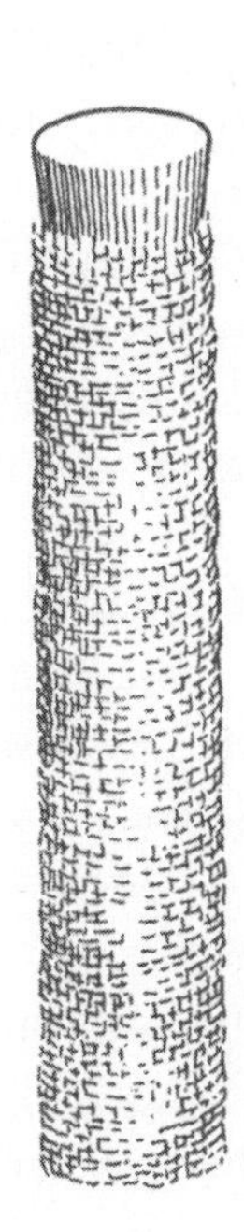

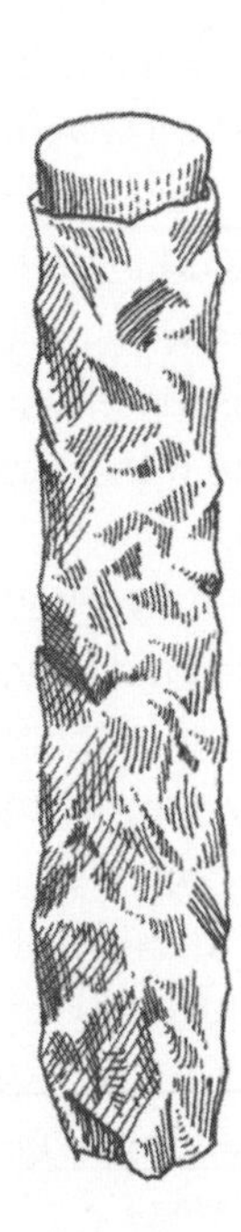

Figure 8.7: Exposing leaves to varying light levels. The first test tube is left uncovered; the second has a single layer of cheesecloth to reduce the light level; the third is wrapped in foil, excluding light. The fourth tube, the control which has no leaf, is not shown.

uncovered, the second was wrapped in a layer of cheesecloth to reduce the light, while the third was wrapped in foil to block out the light completely. To add some variation to the results, several different kinds of terrestrial plants were used. Also, some of the groups used water plants instead of land plants. By adding just enough bicarbonate indicator solution to color the water at the bottom of the tube, no harm was done to the water plants.

Once everything was set up, the groups recorded the time and turned on their bench light. Each tube was gently shaken every five minutes, and notes were made of any color changes in the indicator solutions, using the control as a comparison. One could anticipate that the tube covered with foil would have the highest levels of carbon dioxide, followed by the one wrapped in cheesecloth, with barely any in the uncovered tube, and none in the control. I have long since forgotten whether these results were obtained—likely they were—but, just recently, I conducted a similar but far simpler version of this experiment. This graphically showed the relationship between light levels and carbon dioxide outputs of plants. All that was needed was two large Mason jars (7 inches, or about 18 cm, tall), a box of matches, and a bunch of spinach from the local supermarket.

After adding about an inch (2.5 cm) of cold water to each jar, I took a handful of leaves, cut off their stems, and packed them, standing upright, into the bottom of the first jar, leaving a gap of about two inches (5 cm) at the top. The water kept the leaves moist. The second jar was left empty, as the control.

To test for carbon dioxide, I introduced a lighted match into each jar, dwelling just long enough to see if the flame was put out. As anticipated, it was not extinguished in either one. After screwing on the lids, I stood the jars side by side on my desk, beside a north-facing window. They were not in direct sunlight, so I turned on a 60-watt table lamp for extra light. Three hours later I tested them again with a lighted match, with the same result as before. Any carbon dioxide inside the jar with the spinach was of insufficient quantity to extinguish the flame. By this time, the afternoon light from the February sun was beginning to fade, so I covered both jars in foil for the evening.

Next morning, after fourteen hours in darkness, I removed the foil from both jars. On testing the one with spinach, the lighted match was immediately extinguished, showing that the leaves had generated much carbon dioxide during cellular respiration. Predictably, the match was not extinguished in the empty jar. Leaving the sealed jars on my desk beneath the lamp, I returned to my writing.

On testing the jars after lunch, the lighted match was again extinguished in the spinach jar. Six hours of photosynthesizing in subdued light was obviously not enough time for the leaves to absorb the carbon dioxide released during fourteen hours of cellular respiration in the dark. At this point, I transferred the jars to a south-facing window, where they were exposed to full sunlight. After three hours in the sunshine I repeated the match test: the flame was not extinguished. The light from the sun, even in the middle of winter, had profoundly increased the rate of photosynthesis.

This simple experiment with supermarket spinach had its surprises for me. The spinach, left over from the previous week and offered on special, was not at its freshest and was looking decidedly dejected by the time I had finished. Regardless, it was still absorbing carbon dioxide at an impressive rate while photosynthesizing in the sun. I was also surprised at the carbon dioxide production from its cellular respiration during the night. Although I had no way of measuring the amount of carbon dioxide generated in the jar during darkness, it was obviously high, as shown by the instantaneous extinction of a burning match, a test I repeated several times with the same result. I look at supermarket produce differently now. I used to view the vegetables the same way as I saw the glazed-eyed fishes and cuts of meat. Now I see the greenery as living plants.

Plants use light energy captured from the sun to manufacture sugar, much of which is converted into starch and stored for later use by the plant. Carbon dioxide and water are absorbed in the process, and oxygen is released. The students in my class conducted experiments to detect the sugar manufactured by plants when exposed to light. They also detected the starch that plants manufacture from sugar during darkness. By the time they had finished experimenting, they understood that plants generate oxygen during photosynthesis, and that they generate carbon dioxide when the carbohydrates they manufactured during photosynthesis are broken down during respiration.

In our class discussions of photosynthesis and respiration, the obvious connection was made between plants, which manufacture food, releasing oxygen and absorbing carbon dioxide, and animals, which consume food, use oxygen, and release carbon dioxide. Plants and animals are mutually dependent, and a fine balance exists in the environment to maintain stability in the numerous and diverse communities of plants and animals. Some idea of the diversity and complexity of these communities—ecology—can be obtained during fieldwork, but this is seldom possible for school students, except on occasional daytrips, which have limited value. One useful alternative available in Nuffield biology was for the students to study the diversity of small animals living in soil communities.

Samples of soil, preferably collected in open areas, along with any covering of leaf litter, were placed in an improvised apparatus referred to as a Tullgren funnel. This is simply a large funnel, supported upright and containing a fine-meshed sieve into which the soil sample is placed. A 60-watt light bulb is suspended above the funnel as a heat source, surrounded by an improvised shade to reduce heat loss and allow for some ventilation. A jar containing diluted lab alcohol is placed beneath the funnel to catch and preserve the animals as they move through the soil to escape from the heat and drop through the sieve.

Students might be surprised by the numbers of animals collected from soil, and by their diversity. Aside from familiar animals, like spiders, centipedes, snails, earthworms, earwigs, beetles, and millipedes, are unfamiliar ones like mites and nematodes. Mites are small eight-legged animals related to spiders, and some could easily be mistaken for a small, plump beetle. The largest ones do not exceed a body size of about a quarter of an inch (5 mm), and many are much smaller. Nematodes are worms, but they are unrelated to the familiar ones found in backyards. The common earthworm, which grows to about 10 inches (25 cm) in length, has a segmented body, something like a Slinky toy, and moves with a concertina-like action. Nematodes, in contrast, have unsegmented bodies and move with a side-to-side wriggling motion. While some free-living nematodes reach body lengths of about two inches (5 cm), most of those found in soil are less than a sixteenth of an inch (2 mm) long, and are best seen beneath a microscope.[3] These microscopic nematodes, along with countless unicellular organisms, including bacteria, inhabit the thin film of water surrounding soil particles.

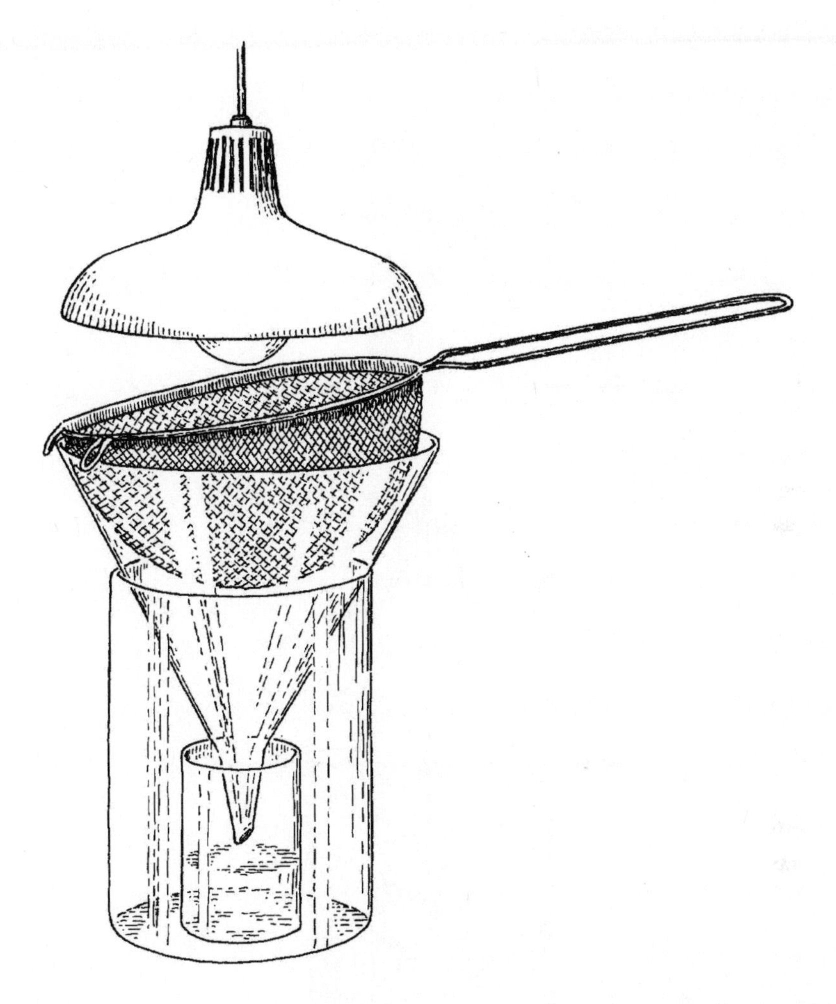

Figure 8.8: A soil sample is placed into the sieve, and any animals escaping from the heat of the lamp fall into the funnel. From there, they drop into the jar containing lab alcohol at the bottom, which fixes and preserves them. This is an improvisation of an apparatus known as a Tullgren funnel.

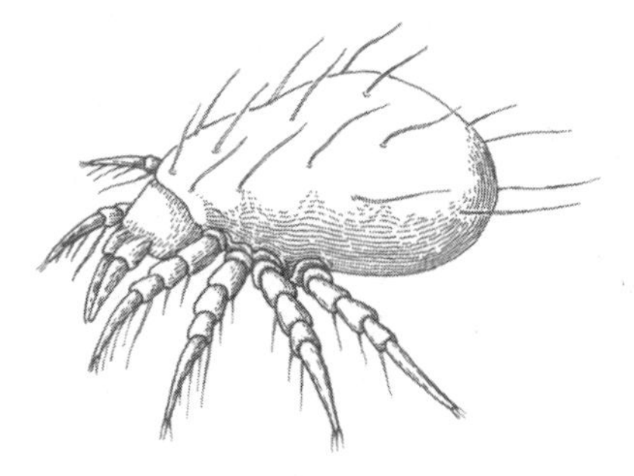

Figure 8.9: This small eight-legged animal is a soil mite.

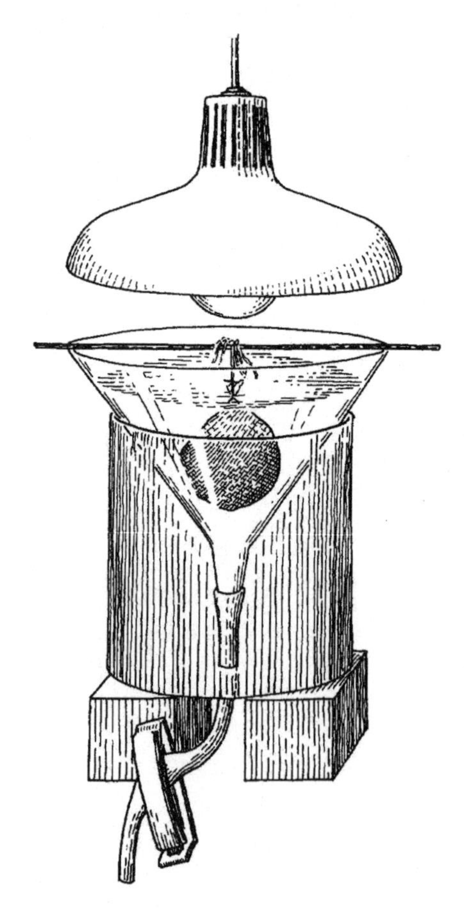

Figure 8.10: The smallest animals can be extracted from a soil sample by enclosing it in makeshift bag of cheesecloth, suspended in a funnel filled with water.

Extracting these microscopic organisms from the soil is done by enclosing a handful of soil in a makeshift cheesecloth bag, which is then suspended in a funnel, filled with water. As in the Tullgren funnel apparatus, an electric light bulb is used as a heat source to drive the living organisms from the soil. The microorganisms are collected by releasing the clamp attached to the rubber tube attached to the stem of the funnel.

If I were teaching biology in school today, I would illustrate the interdependence of plants and animals by introducing students to the intriguing world of plankton. The word *plankton* comes from the Greek for *wandering*, an apt name for the myriad of organisms that drift in the upper layers of ponds, lakes, and seas, where light levels are high enough to support photosynthesis. Most planktonic organisms are small and the plants, collectively referred to as phytoplankton, are the smallest, ranging in size from about one millimeter to less than the size of a human red blood cell.[4] The phytoplankton are the start of the food web. The animals, in contrast, collectively referred to as zooplankton, are larger and most can be seen with the unaided eye.

Plankton is usually collected by towing a special net behind a slow-moving boat. Constructed of fine mesh, these conical nets are typically about six feet (2 m) long, gently tapering, and ending in a cylindrical bottle in which the plankton accumulates. Collecting and examining plankton was an integral part of the marine biology field course I used to teach on the east coast most years. This was one of the highlights for the undergraduates, especially on night cruises where we saw bioluminescence in the plankton.

The most conspicuous organisms we saw in marine plankton samples were the copepods, flea-sized (0.5–2.0 mm) crustaceans with torpedo-shaped bodies and prominent antennae.[5] Most copepods are herbivorous and graze on phytoplankton, the most abundant members of which are the diatoms. Diatoms come in an overwhelming array of shapes and sizes. Some, like the ones shown here, resemble miniature Petri dishes. The smallest diatoms are little larger than human red blood cells while the largest are about one-tenth of a millimeter in size.

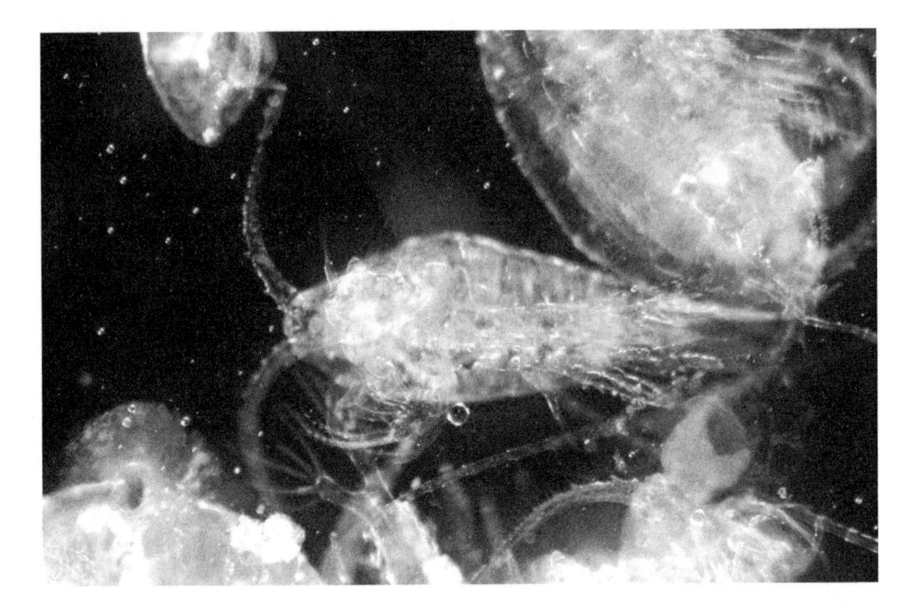

Figure 8.11: These small (1/16 inch or 2 mm) crustaceans, called copepods, are usually the most prominent organisms seen in marine plankton samples.

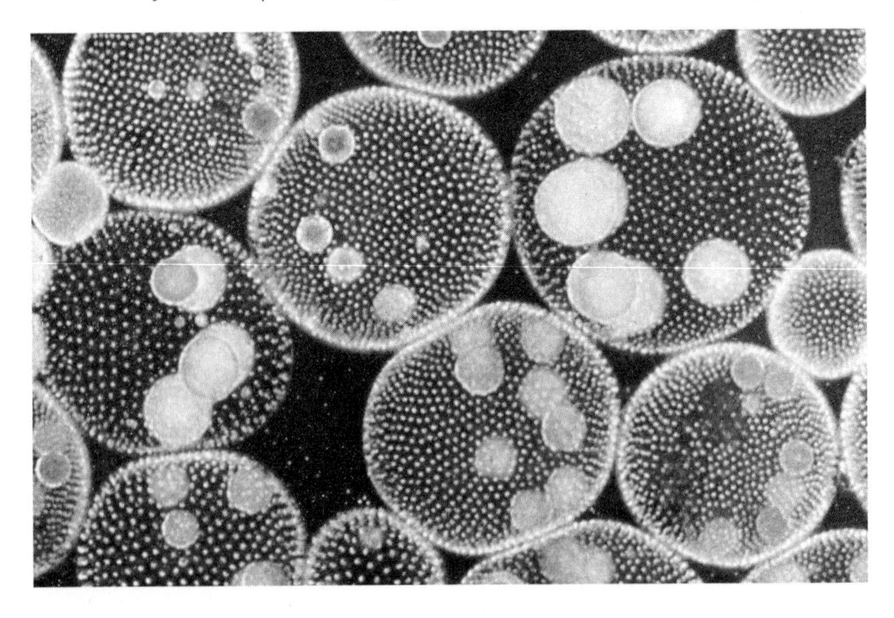

Figure 8:12: Diatoms are microscopic members of the phytoplankton that occur in both fresh and salt water. They comprise a major part of the marine phytoplankton.

I still go collecting plankton, mostly with my grandchildren at a local pond, or at the lake beside my friend's cottage. For this I use a homemade plankton net. All that is needed is a straightened-out wire coat hanger that has been bent into a circle, with the ends twisted together to form a short handle about two inches (5 cm) long. This provides a framework for stretching a nylon stocking over. I used to use the regular full-length ones, but I now use the knee-high variety, which are the ideal length. Once the stocking has been stitched to the frame, all that remains is to bind the short handle to a long pole.

The best way of using the net is to make long sweeps by walking slowly along the water's edge, keeping the net a foot (30 cm) or so beneath the surface. The return stroke is initiated by a broad curving sweep, maintaining momentum so the toe-end of the stocking continues streaming out behind. Depending on the time of year and local conditions, fifteen minutes or so is usually sufficient time to collect a good sample. Aiming the net skyward at the end of the last sweep allows the water to drain away without loss of contents.

Most of the material accumulated at the end of the net often comprises water plants, but this is usually teeming with plankton. The contents are transferred to a partly filled jar of pond water by turning the net inside out while pinching off the swollen end. Much plankton will remain adhering to the inside of the toe-end and this is transferred by touching it against the surface of the water and gently agitating. Holding the jar up to the light usually reveals a darting display of zooplankton, primarily the crustacean *Daphnia*, the common water flea.

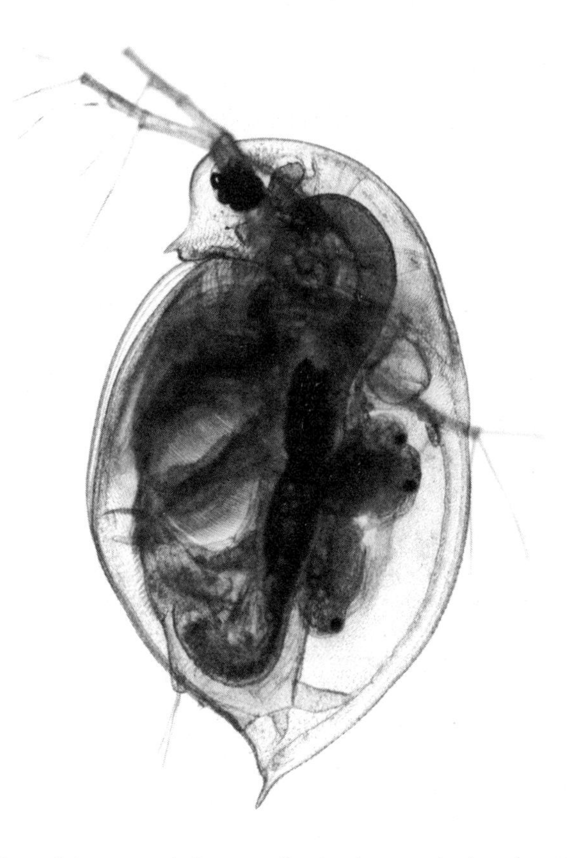

Figure 8.13: One of the most obvious small animals seen darting about inside a jar of freshwater plankton is *Daphnia*, the common water flea. They are crustaceans, like shrimps and lobsters.

A couple of summers ago, I was at my friend's cottage enjoying a large gathering of family and friends. Many youngsters were present, including an eleven-year-old lad visiting from New Zealand. To entertain the youngsters, I collected some plankton with them, using a homemade net. My grandchildren had seen plankton before and were interested to see more, but this was the first time for our young friend from New Zealand. The look of sheer wonder on his face as he peered down the microscope was a joy to behold, and an entire afternoon was spent peeking at plankton. Some of the adults were as enthusiastic as the youngsters. Indeed, it would be difficult not to be beguiled if you had never seen live plankton before. A whole new world in miniature exists beneath the lens.

CHAPTER 9

THE WAY FORWARD

Adults often underrate children, especially the very young. Our grandchildren, above all others, have shown me the error of this. I share my interest in science with them, and one grandson in particular is remarkably intuitive, as the following example illustrates. Without any prior discussion about the blood vascular system, I was sitting next to my grandson, holding a glass in my hand that was resting on my leg, and obviously close to my femoral artery. I told him that I could take my pulse by looking down into the glass—how was that so? He explained that my heart was pushing the blood through my body and that I was picking this up by the pulsing of the blood vessel in my leg, which was causing the glass to make ripples in my drink. He was only eight at the time.

Children are naturally inquisitive, but that window of curiosity does not stay open for very long. Without nurture, the leaf shrivels on the vine. My grandson's interest in science will soon be snuffed out by the way science is being taught in our schools. I do not want that to happen. Nor do I want to envisage a scientifically illiterate world, especially with the serious problems we face in our twenty-first century. What needs to be done to return real science to the classroom?

When I began writing this book, I knew that school science was in disrepair from my experiences with the science curriculum in Canada. I saw similar problems south of the border, but I thought that at least the United States was trying to do something about it with its implementation of the Next Generation Science Standards. Much impressed by the interest shown in the NGSS by states across the country, I started scanning the document, wondering why Canada had not undertaken a similar initiative.

Aside from the inclusion of engineering and technology alongside science, my objections to which have already been given (chapter 5), I was initially impressed. The goal that all students should attain scientific literacy coincided with my own conviction, and some things made good sense.

These included covering less ground in greater detail and building upon the students' previous knowledge, from one year to the next. While the writing style made for some challenging reading in the appendices and other explanatory sections, I thought this was attributable to its having been written by a large committee. A great deal of effort had obviously been invested in the project, and I thought something useful could come from it all.

It was only when I delved more deeply into the NGSS, and especially into the *Framework* document upon which it was founded, that I discovered that educationalists, not science teachers and certainly not scientists, were the driving force behind the new direction of science education. The influence wielded by educationalists, whose works are cited throughout the *Framework* and NGSS documents, came as a complete surprise. So too did the fact that their often absurd assertions appear to pass unchallenged.

My initial thought was that something useful could be salvaged from the NGSS, but I can no longer see how. Aside from the unrealistic goals set for students, and the coverage of so many different subjects, is the overemphasis on the methodologies of science. All of this attention to detail is at the expense of engaging students in science through hands-on practical experience.

I am reminded of something I witnessed a couple of years ago, at a local skating rink. A young lad of six or seven was on the ice with his father. Both had hockey sticks, and the youngster was practicing controlling the puck and firing off shots at an imaginary goal. He would have been happy to enjoy playing at hockey, but his father was intent on drilling him on the finer points of the game. Time and again, he made him repeat a particular move—maybe the father had aspired to play professionally when he was young. The distraught look on the boy's face said it all. If only he could be allowed to enjoy himself on the ice, with his hockey stick and puck, he would learn the details of the game along the way.

Far from achieving scientific literacy, I believe that implementation of the new standards will turn students away from science. And how will teachers cope with teaching everything from the "compositional elements of stars" to the "iterative process of testing the most promising [design] solutions?"[1]

The first step in returning real science to school classrooms is the removal of both engineering and technology from the science curriculum, along with relating science to society. If engineering and technology are considered so important because of their relevance to today's world, they should be taught as separate subjects. Science should also be restricted to biology, physics, and chemistry. The inclusion of space science has always struck me as bizarre, especially given the challenges of conducting experiments in the subject. And while earth science, which includes geology, has relevance to our twenty-first-century problems, the content is far too extensive to be crammed into the science curriculum. Students should leave school with a solid knowledge of basic science, not a flimsy knowledge of all its branches. My knowledge of geology, for example, was nonexistent before I received my first degree, and had absolutely no bearing on my training as a scientist. However, without a good foundation in basic science, I would not have been accepted into a bachelor's degree program in the first place.

While students would still learn about the scientific method, the emphasis would be upon learning through practical experience—the same way that Louis Pasteur made his landmark discoveries—without all the theorizing over hypotheses and testing. For me, learning through practical experience is the only way to teach science, but others would disagree, as the title of one paper in an educational journal appears to suggest.

The paper in question, "Does Practical Work Really Work? A Study of the Effectiveness of Practical Work as a Teaching and Learning Method in School Science," was written by two British educationalists, both former school physics teachers, one of whom has a doctorate in education, the other in medical physics.[2] In the introduction, the authors write that

> Despite the widespread use of practical work as a teaching and learning strategy in school science, and the commonly expressed view that increasing its amount would improve science education, some science educators have raised questions about its effectiveness.[3]

The study explored the effectiveness of practical work by observing twenty-five science lessons involving practicals, in eight schools in England. To determine the effectiveness of practical work, the researchers broke the task down into answering two questions: how well did the students, aged eleven to eighteen, perform the practicals, and what did they learn from them. The practicals covered a wide range of investigations, which included: seeing if there were any differences between heart rate and pulse rate; comparing heat absorption in tin cans of different colors; and measuring current flows in various electrical circuits. The teachers running the various classes were interviewed by the researchers after the lessons, as were some of their students.

In most cases it was found that the students carried out the practicals the way the teachers had intended:

> The practical work observed was, in most cases, effective in enabling the majority of students to do what the teacher intended with the objects provided.[4]

Much of this was attributed to the teachers having provided explicit instructions, and also to their focusing so much attention on the students completing the practicals before the lesson ended. Indeed, the researchers commented that

> The overwhelming sense, from the set of lessons observed, was that a high priority for teachers is ensuring that the majority of students can produce the intended phenomenon, and collect the intended data.[5]

In marked contrast, there was

> considerably less evidence that they were as effective in getting the students to think about . . . the ideas that were . . . intended by the teacher.[6]

Remarkably, many of the students were unfamiliar with the subject matter of the practicals, and most of the teachers offered little or no explanations of the underlying science during the lesson. Instead, the teachers' attention was focused on the students carrying out the practical, as instructed, before the lesson ended. For example, the teacher dealing with pulse rate had chosen

> not to discuss the circulatory system before they began, explaining when interviewed that she believed the connection between heart rate and pulse rate would emerge from the data.[7]

When the results of the pulse rates and heart rates were written up on the board, the teacher asked the class whether the rates were the same. Because the students were unfamiliar with how blood circulates through the heart and around the body, they did not know that they should be identical. Some of the students responded that the rates were different, but the teacher's response was dismissive: "near enough," she replied, to which someone called out, "But 106 and 90 are miles apart."[8]

The teacher of the class experimenting with heat absorption by different colored cans was equally ineffective. His primary objective was for the students to finish the practical with the expected results. To this end, he used everyday language in his explanation of how to conduct the investigation. After initially introducing the term *heat*, he made no further references to scientific terms, or to any scientific principles, throughout the entire lesson.

Only one of the twenty-three[9] teachers participating in the study gave a substantial account of the underlying science behind his practical, which was an investigation into the relationship between current and voltage.

The authors summarized their findings:

> Our observations of these 25 lessons suggested that the practical tasks used were generally ineffective in helping students to see the task from a scientific perspective, and to use theoretical ideas as a framework within which their actions made sense or as a guide to interpreting their observations.[10]

Clearly, the reason why the practicals were so ineffective in teaching the students about science was because of the negligence of the teachers. It came as no surprise to me to learn that two-thirds of them were teaching outside of their areas of specialization.[11] A biology teacher, for example, might be supervising a chemistry practical.

The understated conclusion of the researchers was that

> Our study suggests that practical work in science could be significantly improved if teachers recognized that explanatory ideas do not "emerge" from observations, no matter how carefully these are guided and constrained.[12]

I find it difficult to understand why teachers would set practicals on subjects that had not yet been explained in class. The entire objective of practicals is for students to learn about particular aspects of science through their own hands-on experience. What is the point of practicals for the sake of practicals? As I write this, I am reminded of all the time we spent in chemistry classes, in the 1950s, carrying out small-scale inorganic qualitative analysis. In spite of its long name, all this entailed was conducting a series of tests on small samples of an unknown solution, provided by the teacher, to determine what it contained. Using miniature test tubes, dropping pipettes, beakers little more than an inch (2.5 cm) high, and a centrifuge to spin-down precipitates, we made our way through a recipe book of tests: "add two drops of concentrated hydrochloric acid . . . saturate the solution with hydrogen sulphide gas . . . conduct a flame test . . . separate off the precipitate that forms by centrifuging . . ." I still have the book.[13] If everything went well, all we had to show for our efforts at the end of the lesson was a list of the chemicals contained in the solution. Aside from imparting some skills in handling laboratory equipment, these practicals taught us very little about chemistry.

Regardless of whether the NGSS is adopted or rejected, school districts across the country will be creating new science curricula. The best way to set about the task would be to follow the Nuffield approach. This would involve small groups of science teachers, with proven ability in the classroom, working in conjunction with a cadre of professional scientists as resource people. Curriculum consultants and educational specialists would be excluded, thereby eliminating their nonscientific obfuscations—along with their consultation fees. I would also exclude faculty members from departments of education, on the grounds that their academic interests are educational, not scientific. Educational administrators would preferably be excluded too, but that is likely an unrealistic aspiration.

To avoid the pitfalls of writing by consensus and producing challenging reading like the *Framework* document, small groups of people with a talent for clear and concise writing would generate the final documents.

Hands-on science would occupy increasing amounts of class time and, by grade seven, students would be spending most of their time conducting experiments themselves. At this point, science would be separated into its three main branches: chemistry, physics, and biology. Ideally, a set of textbooks would be published for high-school students, one for each grade, like those in the Nuffield Biology series, with a companion series of teachers' guides.

My dismissiveness of educationalists, here and elsewhere, may strike some readers as extreme, but I make no apologies for drawing attention to the damage done by so many of them to the way science is taught in our schools. And their destructive reach extends beyond the school classroom, as I discovered while nearing completion of this chapter. I was as surprised by what I learned, as you might be when you read what follows.

Through my interests in animals, both living and extinct, I know a number of researchers across North America and beyond. Some time before writing this chapter, I was chatting on the phone with a good friend, a biology professor at a university in the United States. In catching up with what we had been doing, I told him about this new book. Life was treating him well, and he sounded calm and relaxed. Jump forward several weeks to our next conversation, and things had completely changed. This became apparent when I happened to mention my concerns about the negative impact educationalists were having on school science. At this, he began telling me of the havoc they were wreaking at his university. Because of educationalists, his teaching and research, along with that of the rest of his department, had been disrupted for weeks.

I learned that universities in the United States, unlike those in Canada, are usually accredited, a process conducted and administered by a number of regional authorities. Once a university is accredited, it has to undergo periodic reaffirmation. While this major and comprehensive process only takes place every ten years or so, universities have to supply their accreditation authority with continuous updates to show they are fulfilling their mandate. The primary objective here is for universities to substantiate that they are providing their students with the best learning available to them. To document how well students are learning, universities have to provide an assessment of the learning outcomes of their students. I have difficulties trying to visualize just what *learning outcomes* are and how they can be accessed, but I am not an educationalist.

Remarkably, there is a National Institute on Learning Outcomes Assessment (NILOA), based at the University of Illinois and Indiana University. Although their website was not helpful in providing a definition of what they assess, I did find a useful one at Brigham Young University. Here, learning outcomes are defined as: "what a student is expected to be able to DO as a result of a learning activity."[14] Given that accredited universities must provide documentation of their learning outcome assessments, I should not have been surprised at the existence of the NILOA, nor of the cottage industry for educationalists that has arisen from these assessments.

And most US universities have an Office of Institutional Assessment to assist departments and faculty members to comply with the requirements.

The Office of Institutional Assessment at my friend's university provides faculty members with a guide to assist them in completing their assessments of their courses. Illustrated with the educationalists' obligatory flowchart, this particular diagram depicted a circle of different colored segments. This was attributed to Dr. Trudy Banta, professor of Higher Education at Indiana University–Purdue University Indianapolis. A pioneer and leading authority in the field of outcomes assessment, Professor Banta has consulted with administrators and faculty in almost every state in the union. In addition to the guide, the assessment office provides a video, adapted from Trudy Banta. With its references to pedagogy, stakeholders, mission statements, rubrics, and the like, it was as impenetrable as all the other educationalists material I have struggled to understand.

My friend also had to produce a Quality Enhancement Plan (QEP), which "describes a carefully designed course of action that addresses a well-defined and focused topic or issue related to enhancing student learning."[15] His frustration at the sheer waste of valuable time in completing these arcane educationalist exercises sounded loud and clear across the phone. And what would be achieved? Will his efforts enhance the learning experiences of his students? Would going through such a process have improved the educational experiences of my former students when I was teaching at university?

Before taking early retirement, I taught a biomechanics course in the Department of Zoology at the University of Toronto. Titled Vertebrate Mechanics, this was essentially an engineering course for zoologists that was focused on locomotion and skeletons. The format of the course was a lecture, followed by a two-hour practical, the latter being the major component. Because of equipment restrictions—we only had six wind tunnels for the aerodynamics section—there was a class limitation of twenty-four students. Most were in the third or fourth year of their BSc degrees, and I got to know them quite well during the practicals. One of the outcomes of the course was a textbook, which I hoped would encourage others to teach similar hands-on courses.[16]

Several years ago I received an email from someone I knew from my museum days. She had recently visited the hospital with a broken bone and wrote to tell me that

> The resident who examined my recent fracture wanted you to know that his choice of occupation as an orthopedic surgeon came about because of your inspiring course in biomechanics. He developed an incurable disease, "bone fever," which he gratefully attributed to your brilliant lectures.

I received a much longer letter, from one of my former students, written thirteen years after she took the course:

> I have been meaning to write to you for some time now just to say thanks. . . . As it turns out, I was unsure which path to take in life and so applied for a Master's . . . as well as for the Veterinary Medicine program . . . and here I sit, a DVM [Doctor of Veterinary Medicine].
>
> I found it to be the single most useful course I took in all my undergraduate years for its relevance to my practice of veterinary medicine. Knowing that one should appreciate bone and skeletal structure and the functional differences of species . . . was extremely useful in both diagnosis of musculoskeletal problems and surgery. I referred to my notes from the course more than from any other course, and spoke to classmates of the example of the saber-toothed tiger skeleton versus the prehistoric cervid [deer] as a description of form relating to function.[17]

My reason for quoting from these letters is not to impress but to support the claim that my course had a positive influence on the students. Could I have enhanced their learning experience? Undoubtedly yes. One way I tried doing this was to be current with the latest research, incorporating any interesting new findings into the lectures. I also introduced new experiments and improvised new pieces of homemade equipment from time to time. Could I have improved their learning experience by investing time deciphering educationalists' texts to devise mission statements, outcome assessments, QEPs, and all the rest? Undoubtedly not. As my friend south of the border said, "Why should we waste our time on this nonsense when we all know it makes absolutely no difference anyway?"

How realistic am I in thinking that new science curricula could be introduced into science classrooms in the United States by following the Nuffield approach? First, could dedicated science teachers, of proven ability in the classroom, be found for the task? I have absolutely no doubt of this from my interactions with teachers in the United States, who were every bit as enthusiastic about engaging youngsters in science by *doing* science as I am.

One of these encounters took place in Chicago in 1999, during a daylong workshop with teachers and their students. This was organized by Project Exploration, an outstandingly innovative nonprofit organization for bringing science and scientists into the lives of youngsters who would otherwise miss such opportunities.[18] I staged a morning of practical demonstrations to show what can be done in classrooms. One of the activities I introduced was determining one's center of gravity by balancing on a plank of wood. This led to investigations into balance, and then into walking on two legs and on four. The teachers participated in the activities with as much enthusiasm as their students. Incidentally, in 2009 President Obama named Project Exploration as a recipient of the Presidential Award for Excellence in Science, Mathematics and Engineering Mentoring.[19]

Aside from being good teachers, prospective participants in creating new curricula would have to be convinced that their time and effort would result in something worthwhile. UK teachers faced the same situation in the sixties, when they wanted to break away from the old chalk-and-talk way of teaching science. Canadian teachers, jaded by the inane science curricula they must now follow, have all reacted positively to my recent demonstrations of what can be achieved with hands-on science, using everyday items. I would not expect to find any different reactions from science teachers in the United States.

The biggest barrier to teachers trying to return real science to the classroom is the opposition they would face from the educationalist and curriculum specialists entrenched in the system. How could educational authorities be convinced of the need for change when their own "experts" on teaching science tell them otherwise? Enlisting the help of the scientific community to authenticate the shortcomings and inaccuracies in science curricula is the obvious solution.

One of the reasons for the success of the Nuffield Science Teaching Project, back in the sixties, was the strong links that were forged with the scientific community.[20] This gave teachers access to specialists, in various areas of science, with whom they could consult during their creation of curricular content. What are the prospects of similar links being formed, during these busier times, in North America? Some of the groundwork for such ties was begun by an initiative from the White House.

On November 23, 2009, President Obama announced the launching of a new project to reinvigorate science education in the nation's schools. Named National Lab Day, the objective was to promote hands-on learning by connecting teachers with scientists who would volunteer to visit classrooms to conduct practicals.[21] Scientists interested in participating would identify their fields of interest when signing up. Teachers would send in their requests for practicals relating to their lesson plans, and the scientists would periodically receive lists of teachers seeking help in areas matching their interests. The program would culminate near the end of each school year with a series of special events, staged in May.

Most of the volunteers were professional scientists, from universities and other research establishments, along with graduate students working on higher degrees. I make the point because there are organizations in Canada, perhaps in the United States too, that seemingly provide similar services. However, most of these volunteers are people with little or no background in science. A few years ago my youngest granddaughter had a "scientist" in her classroom. As it happens, he was a fireman by profession and, judging from the handout she brought home, he had some shortcomings in his science.

Much impressed by National Lab Day, I regretted that we did not have a similar program in Canada. I signed up in January 2010 and may have been one of the first, if not the only Canadian to do so. When I entered my alien zip code during the registration process, it was interpreted as being in New York, the state across the water from Lake Ontario. Consequently, I only received requests from schools in New York State. As I explained to the teachers with whom I interacted, I would have loved to have paid a classroom visit, but we had to make do with giving and receiving assistance remotely. Nevertheless, it all worked out very well, and I received some appreciative and enthusiastic emails from teachers I was able to help. The subjects ranged from human respiration to geothermal heat.

Unfortunately, the requests petered out within a few months of my joining, and I received no reply to my email enquiring why. I assumed that living north of the border was the likely cause and did not pursue the matter further. It was only much later, while searching the Internet for news of the project for this book, that I discovered its apparent demise. Regardless, there are still websites that refer to the project. The American Chemical Society, for example, has a site where members are urged to sign up, but clicking the National Lab Day tab eventually returns with the "This site can't be reached" message.[22] I received the same message when I attempted visiting the National Lab Day site that I accessed when I joined.[23] Then I discovered that the Department of Energy has a website, dated as recently as July 10, 2015, announcing that it "hosted the second in its series of five National Lab Days on Capitol Hill."[24] Had National Lab Day ceased to exist or not?

When I found a website for the National Lab Network and read that it was "a national initiative that connects K–12 teachers with science . . . professionals to bring hands-on learning experiences in all 50 states"[25] I felt optimistic. The same six sponsors who had supported National Lab Day were acknowledged, so perhaps this was its reincarnation. This was confirmed by a posting on Facebook, dated February 15, 2011, noting the official change of name.[26]

The website of the National Lab Network appears to be fully operational with its changing lists of projects that scan across the bottom of the screen, along with changing images of national leaders who support the program. However, none of the tabs are functional, including the ones for volunteers and teachers to use for signing up. I can only assume that this initiative has gone the same way as its predecessor. This would seem to be confirmed by the absence of any recent postings on Facebook.

Without any information on the reason for the ending of operations, I would surmise it was because of the difficulties of finding enough volunteers from academia to keep up with the requests. This does not surprise me. When I joined the Royal Ontario Museum, I could scarcely believe the number of letters and phone calls I received from teachers requesting classroom visits—and that was back in the sixties. Mindful of how much time would be lost from my working week if I acquiesced to the requests, I decided to decline them all.

Does this mean that science teachers would be unable to draw upon the expertise of scientists when devising new curricula for the classroom? I believe the answer is likely to be yes, if they were relying upon academics who had classes to teach, papers to write, projects to complete, grant proposals to prepare, and all the rest. However, if *retired* scientists were approached, the situation would be entirely different.

While some scientists continue with their research after retirement, others direct their inquiring minds elsewhere. And there are doubtless those who face the end of their working careers with the same trepidation as so many other people. I am reminded of the banker I used to chat with at the bus stop most mornings. His seniority was apparent from the overseas trips he used to take, and financial insecurity was obviously not one of his problems. One morning, out of the blue, he announced he was about to retire. When I offered my congratulations, he became quite distraught, asking me what on earth he was going to do with all that time on his hands. My conciliatory suggestions about enjoying his interests and pastimes revealed that he had absolutely nothing outside of banking.

If an organization like National Lab Day made an appeal for retired scientists to be part of a science resource for teachers preparing curricula, I am sure there would be no shortage of volunteers. This would be especially true if they were made aware that science, as they know it, is in jeopardy in our schools because of the dictates of scientifically illiterate educationalists.

Before taking early retirement, I had neither time nor motivation to examine school science curricula. How quickly things changed. After discovering the serious state of school science, I asked my former colleagues if they had ever taken the time to look at the senseless science curriculum. None of them had. However, from some of the things they were saying about the fresh intake of students into their university classes, they were not surprised at what I told them.

Unlike the banker at the bus stop, I am as busy now as I was before I retired. Much of my time during the last few years has been taken up with my concerns over what has been happening in science classrooms across North America. If an organization along the lines of National Lab Day was initiated to connect retired scientists with schoolteachers striving to return science to the classroom, I might be able to offer some useful contributions.

CHAPTER 10

A CLASSROOM CHARTER

Most of the focus of this book has been upon what science is and how it should be taught, but there are some additional issues to consider beside curricular content and teaching tactics. While these issues have direct relevance to the teaching of science, they also apply to other subjects too. Some would argue that the two hotly debated items dealt with here—how students are assessed and how they behave—belong in an educational tome, not one centered upon science. My counter to this is that science cannot coexist in classrooms where these critical issues have not been resolved. Imagine, for example, a teacher trying to conduct a hands-on science class where some of the students are more focused on misbehaving than on handling equipment.

There are those who would say that if science is made sufficiently interesting and engaging, misbehavior would not be an issue. I suspect this idealistic view would not be shared by those who have taught in school, especially not to classes of boisterous youngsters.

Some educationalists advocate that learning should be fun, almost as if entertaining students is part of a teacher's job. There were certainly some moments of fun when I was school teaching, like the time the fifteen-year-old boys in my class were dissecting fresh cows' hearts. All of them had their fingers bloodied and red from probing the large blood vessels leading into the various chambers of the heart. When Melvin, the class comedian, made a comment about one of the other boys being okay except for picking his nose—the implication being that excessive nose-picking had been the cause of that boy's bloodied fingers—we all enjoyed the joke.

Dissecting animals, and their organs, certainly kept my students engaged, but there were many mundane practicals too, like monitoring a test tube containing a snail for half an hour to record color changes in the liquid at the bottom. Did they react any differently? Absolutely not. Bad behavior has little or nothing to do with what is being taught, or how much fun students are having. However, classroom behavior, along with how students are assessed, has a considerable bearing on how well students learn science, as we will see.

Science is an exacting subject, and students should strive to reach the highest levels of competence. When conducting experiments, for example, care has to be taken to obtain measurements as accurately as possible. These must be carefully recorded, along with an account of what is done, in sufficient detail that others can repeat the experiment. Keeping a good notebook in the lab is vital; the progress of science depends on following what has been done before.

I recall a young lad in one of my classes being concerned about accidentally spilling some of the reagent used in his experiment onto his notebook. I told him not to worry because the stain on the page showed that the notes had been taken at the same time as the experiment. As I reiterated to the class, a lab-weary notebook was a good sign that the book had been used for its intended purpose. Incidentally, it is because of chance spills and splashes in the lab—mostly with water—that pencils are always used, never pens, something I learned from one of my science teachers. This is especially important for field notebooks; they can be dunked in the sea without any loss of data.

Keeping proper lab notes is part of the training, and this can only be taught in a properly regulated environment. It would be impossible to instill the ethos of care and attention, so essential to science, in a disruptive classroom. Inattention is the antithesis of the scientific method; so too is the departure from rigor. The teacher conducting the class practical comparing heart rate and pulse rate was wrong to say the results obtained were "near enough," and the student was perfectly correct in challenging her assertion (see chapter 9). High performance standards have to be maintained in the science classroom, just as high standards of cleanliness have to be maintained in the operating room. This ethic is engendered by students completing assignments to the best of their ability: "near enough" is not good enough. I therefore make no apologies for laboring this issue in the section that follows.

In this age of entitlement, students expect, and likely receive, passing grades regardless of whether they are deserved. Giving low grades is anathema to some educationalists. Many school authorities, both in Canada and the United States, forbid teachers from failing students. In 2012, Lynden

Dorval, a high-school physics teacher in Alberta with thirty-five years of teaching experience, was dismissed for giving zero grades to students who had not completed their assignments.[1] This was because his action contravened the local school board's policy of not giving zero grades.

The rationale for the no-zero policy was based on the findings of a study undertaken by a team of researchers from the faculties of education at three Alberta universities. Extending over a period of two years, the findings were published in 2009. The lengthy report described the investigation as being of a

> multi-stage mixed-method design that consisted of four main stages: Stage One—Literature review and lecture series, Stage Two—Focus groups with educational stakeholders, Stage Three—School-based data collection with students, parents, and educators, and Stage Four—Data processing, analyses and reporting.[2]

The publications referred to in stage one included the names of two educationalists, Grant Wiggins and Jay McTighe, whose work has been mentioned earlier, in chapter 1. As these are the only authorities listed whose publication I have consulted, I am in no position to comment on the rest of the literature reviewed in the study. Given that most appear to be authored by educationalists, I suspect I would gain little from attempting to read them. However, I can comment on one important aspect of the third stage of the study, namely the school-based data collection. It is clear from the report that the students who participated in this part of the study had the greatest influence on Alberta's decision to banish zero grades:

> Classroom-based assessment was . . . perceived to be a form of reward and punishment that teachers deploy at their discretion. The bulk of the comments associated with this perception come from students, and most of these were negative toward the idea of grades reflecting behaviors. Both elementary and secondary students voiced displeasure at teachers assigning zeros and, as one secondary student suggested, it is "unfair to get zero on homework not done" and an elementary student stated, "Zeros are cruel."[3]

The implication here is that it is the *behavior* of not completing an assignment that is being punished with a zero grade, as opposed to being given a zero for an assignment that has been poorly executed. If you too are having difficulty with this concept, Michael Zwaagstra, a high-school social studies teacher and author of the aptly named book, *What's Wrong with Our Schools and How We Can Fix Them*, can help:

> The philosophy underlying no-zero policies is quite simple. Proponents believe teachers should always separate behaviour from achievement when grading students. Since cheating on tests, handing in late work, and refusing to submit assignments are all examples of behaviour, they should not affect students' academic grades. Instead, they argue, teachers should correct poor behavior in other ways.[4]

The students in the Alberta study complained about losing marks for lateness and incomplete work as well as for work not done at all:

> Elementary students expressed their dislike of having "marks taken off for late work" . . . and a secondary student didn't want to "lose marks for incomplete homework." Another student said, "Once you get parents involved teachers lighten up."[5]

The last quote could have been made by the son or daughter of the parent who informed the researchers that she was "not at all OK with late assignment penalties."[6]

The last student quote highlights how adept youngsters can be at exploiting situations to their own advantage. The influence of the students in this study is apparent from the following recommendations that were made for Alberta's schools:

> Student voice must help shape assessment practices.
> Students will engage in assessment design.
> Late assignments are accepted without penalty.
> No-zero policies support student-learning outcomes.[7]

In terms of the roles of student and teacher, "student decision making"[8] included students having the following responsibilities:

Prioritize tasks and activities.
Decide on preferred learning and work styles.
Manage time.[9]

While "teacher decision making"[10] included the recommendation that they

Involve students in decisions about their learning.[11]

The concept of students making decisions about their learning of science is complete nonsense. Having seen examples of gross negligence and lost learning opportunities by teachers running classes outside their own field (see chapter 9), one can only imagine what would happen if students played a part too. It makes equally bad sense for students to "engage in assessment design."

Following his dismissal, Lynden Dorval was hailed as a hero by the media for standing up for his teaching principles. In one interview, the physics teacher summarized how the no-zero policy lowered standards in the classroom: "It's a way of pushing kids through, what the Americans call 'social promotion,' and making the stats look good."[12]

Social promotion is where students are promoted to the next grade at the end of the school year in spite of their unsatisfactory academic performance. The rationale for the practice is to protect the individual's self-esteem and encourage socialization by age. The elimination of zeros also inflates marks, giving a false impression of the academic achievements of the student as well as that of the school.

Dorval was soon hired by a private school, a traditional establishment where students expected zeros for missing assignments. Two years later, after he had retired from teaching, an Alberta appeal tribunal ruled that he had been unfairly dismissed and compensated him for his financial losses.[13]

Gina Caneva, a high-school English teacher in Chicago, has a story to tell about the no-zero policy and how it affected her and the students she taught.[14] A decade ago, she was one of the founding teachers at a small new high school in the Chicago Public Schools system, the third-largest school district in the United States. For a young teacher in a new school, her first year was particularly demanding. Her primary problem was the low academic standards and work habits of so many of her students. Aside from not handing in homework, some students failed to complete assignments even during class time. She remembers one student in particular for what she did during a final exam. After reading through the questions and finding she was so unprepared, the student said she would take the exam the next day. When her teacher explained she could not do this as it would be unfair to her classmates, she walked out, saying she was going to fail anyway.

During the first year of operations at the new school, students were given zeros for not handing in assignments. Gina Caneva noticed the impact this had, and how many of the students tried to catch up by earning partial credits in voluntary tutoring sessions during lunchtimes and after school. Even some of the first-year students who were not expected to graduate from high school signed up for summer school, to improve their chances. For readers not familiar with the US school system, such unpromising youngsters are categorized as off-track students. To be on track, students in their first year of high school must earn at least five full-year course credits, and no more than one F (failing) grade in a core course subject, which includes English, mathematics, science, and social science.[15] Regardless of all these good efforts, by the end of the school year only 59 percent of the students were considered on-track for graduation. What could be done to improve the situation?

The principal proposed that the school adopt a no-zero policy, on the grounds that a zero could pull a student's grade down so far that recovery was unlikely. To persuade teachers this was the right option, she had them read an educational article. There, the argument was made that the traditional grading system unfairly penalized students at the lower end of the academic scale.

The standard grading system in schools is for the passing grades, designated A to D, to be divided into ten percentage-point intervals: A, 100–90; B, 89–80; C, 79–70; and D, 69–60. Anything less than 60 percent is given a failing grade of F. According to the argument in the article, as the F grade encompasses a 59 percentage-point interval, in contrast to the ten-point intervals of the passing grades, it is unfair to the weaker students. One option to offset this seemingly unfair disparity was to record failing grades as being no lower than 50 percent. A student who did especially poorly in an assignment, with an actual mark of zero or thereabouts, would therefore be in a better position to recover from the setback.

When the principal put the no-zero proposal to the vote, the majority of teachers decided in favor, thinking the policy could always be changed later if it were not working properly. Some things continued the same way as before. Students who did little work in class, failed to hand in homework, and opted out of exams, continued doing the same things. However, there was a major change in those who had worked hard to improve their grades by attending tutoring sessions—they stopped going. Why should they bother now that they had a safety net? On the other hand, there was a significant rise in the on-track rate among the first-year students because the laggards, who would have received zeros before, were now scoring 50 percents.

Gina Caneva, along with some other teachers, wanted to reverse the detrimental policy, but the administration refused. Why should they do otherwise when the on-track rate had increased from 59 to 87 percent, far exceeding the Chicago Public School average?

Moving on, Gina Caneva now teaches at a selective enrollment high school in Chicago. She eloquently encapsulates the folly of the no-zero policy:

> It is incredibly difficult for people in our country to claw their way out of generational poverty. A good education is central to that struggle. And yet we are saying to young people in Chicago who have grown up in the deepest poverty, "You don't have to work hard to pass. You can miss half of your assignments in all of your classes, and you can still graduate from our high school."[16]

In spite of—actually I am cynical enough to say because of—the senselessness of the no-zero policy, school administrations across the United States have been incorporating the practice into their grading policies. In Georgia, for example, the Lowndes County Board of Education has had a no-zero policy in place for several years. According to former superintendent Dr. Steven Smith,

> Failure is unacceptable. If a student fails, we must determine why and seek a solution to achieve success for the student. Thus evaluation, re-teaching, and re-assessment of students who fail to master the skill the first time is vital.[17]

The idea that a teacher should have to spend time and effort evaluating the reason why a student handed in a poor assignment, for example, is absurd. Equally ridiculous is the notion that a student's failing in some area should require the teacher to re-teach the material and re-assess the student. Here, the responsibility for failure is placed squarely on the teacher rather than the student.

Dr. Smith, the former superintendent, who holds a doctorate in educational administration and supervision, started off as a schoolteacher. Had he forgotten how demanding teaching can be? How could teachers possibly find time to repeat material for every inattentive student in their class? And why should they? Do students bear no responsibilities for paying attention in class and asking questions as problems arise? With educational policies like this, it is little wonder that the sense of entitlement has become so entrenched among students of all ages.

Aside from eliminating zeros, Fairfax County Public Schools, in Virginia, has considered allowing "students across the county to receive credit for submitting corrected answers to questions they got wrong on tests."[18] Gosh, before long students will be setting their own tests and doing all the marking! Sarcasm aside, what message would this convey to a science class? Do not worry about measurement errors, setting up your equipment properly, and recording all the data; if you make mistakes you can just repeat the experiment some other time. This is an unsound foundation for a career in science.

One of the experts named in the *Washington Post* article where the above story was reported is a former Toronto high-school teacher who has written three books on grading. He holds very strong views on zero grades:

> Education consultant and grading expert Ken O'Conner said . . . that giving students a zero is "morally and ethically wrong. As soon as a kid gets even one zero, they have no chance of success."[19]

But failure is a normal part of the learning process that has been with us since early infancy. Do toddlers stop trying to walk after their first fall? We likely learn more from our failures than from our successes. Failure is the spur that drives us on, the determination to avoid repeating the same mistake. Sir James Dyson, British inventor of the cyclone vacuum cleaner, said, "You never learn from success, but you do learn from failure."[20] He went on to say that, "You don't have to bother to be creative if the first time you do something, it works."[21]

Failing my eleven-plus exam was the biggest academic failure of my life. However, I soon recovered, and I suspect this was the impetus that rocketed me from the bottom of my junior form to the top of the senior class. And now educationalists are telling us that failure spells disaster.

No-zero grading policies are championed by consultants and school administrators, but they do not sit well with most teachers. During a presentation of the new grading policies at a meeting of the Fairfax School Board, the result of a teacher survey was announced: 65 percent of teachers believed in giving zeros for work not handed in.[22]

Regardless of the opposition of teachers to the no-zero policy, it appears to be taking hold across the United States. This is apparent from some of the banner headlines appearing in the press:

"No-Zero Policy to Be Implemented in South Carolina Schools"[23]

"Inflating Grades Lowers the Bar in Maryland Education"[24]

"Alabama School Implements 'No Zeros' Grading Policy"[25]

"Tucson Teachers Tell Principal Her No-Zero Grading Policy Is a No-Go"[26]

"Jeb Bush Says Orange County Schools Made It Impossible for Students to Receive Below a 50"[27]

One of the things that struck me when I first walked into my grandchildren's primary school was the inspirational messages that appeared on corridor walls and in classrooms and halls: Believe in Yourself; Every Expert Was Once a Beginner; Be a Voice and Not an Echo; and the like. The rationale for them seems to be that youngsters need constant encouragement to boost their morale. On one school visit, I recall, there was a small awards ceremony for the first-graders, with certificates and congratulatory handshakes for being attentive, punctuality, good listening, and other positive attributes.

When I was a teacher, such things were taken as givens, part of the expected conduct in class, and rewards were neither expected nor conferred.

Behavioral problems were likely not an issue at my grandchildren's school, but what was the situation in middle and high schools? My first experience with older students in North America was back in the early seventies, soon after my arrival in Canada. One of the senior administrators at the museum where I then worked asked if I would consider visiting his daughter's class, to tell them about dinosaurs. He had always been most helpful to me, and I was more than happy to oblige. The young teenagers in the class were polite and well-behaved, but I was surprised at the absence of complete silence when I began to speak. Reminding them that only one voice should be heard at a time, I said that if one of them wanted to take over, I would be happy to sit down and listen. They got the point.

During the last few years, I have visited several middle- and high-school classrooms around where I live in Ontario, some thirty miles north of Toronto. I have been impressed by what I have seen, both in the behavior of the students, and in the good relationships between teachers and students. Unfortunately, this is not the case in so much of North America and beyond, and I have been surprised, even shocked, by some of the things I have learned. The abysmal student behavior in one particular part of the world took me by complete surprise, but I will begin in the United States.

So much of the behavior seen in classrooms is determined by the location of the school but, as will be seen later, there are some important and noteworthy exceptions. Inner-city schools in the United States have suffered some of the worst extremes of student misconduct, often related to criminal activities involving drugs and firearms. The severity of the problem led to the Drug-Free Schools and Communities legislation of the 1980s[28] and the Gun-Free Schools Act that followed.[29] This legislation was the impetus for the zero-tolerance policies that have been adopted by school authorities across the country. Here, offenders who bring banned items into the classroom, like guns or drugs, are automatically removed and punished without being given the opportunity to explain any extenuating circumstances.

The zero-tolerance policies of schools have been extended beyond drugs and weapons to include alcohol possession and many other misdemeanors, including threats of violence, fighting, and even dress-code infractions. School principals are obliged to take action whenever the policy is broken. While this makes sense, some school authorities prohibit teachers from using their own discretion. This has led to students being disciplined for the most ridiculous of perceived offences, from the possession of cough drops and mouthwash to nail files and water pistols. One ten-year-old was expelled after she realized her mother had put a small knife in her lunch box, even though she reported the discovery to her teacher, who commended her for her action.[30]

Remarkably, some schools use security officers to enforce law and order, which sometimes leads to lawsuits. An attorney in New Mexico describes a typical scenario:

> A kid who does not want to transition from one activity to another may throw a book or push a desk. . . . The teacher then calls security. In some cases, the room will be cleared of other students, leaving three or four adults to surround the kid who is considered "noncompliant." If the kid ends up on the floor, it can escalate into head-banging or thrashing, and may result in injuries.[31]

The lack of common sense in the implementation of the zero-tolerance policy has been responsible for an inordinate number of students being suspended and expelled from school. It has been reported that since 2009 the national average has risen to ten percent, the highest it has ever been.[32] This amounts to a staggering number of students being suspended and expelled from public schools each year: well over three million in 2006, according to the US Department of Education.[33]

Ridding the class of disruptive students allows the others to continue their work, while the removal of criminal elements makes for a safer school environment. However, unnecessary harm is done to those expelled unjustly, or for trivial reasons. Furthermore, giving troublemakers an unscheduled break from school is more of a reward than a punishment, and provides them with opportunities for getting into further mischief.

The perception that more harm than good has come from the zero-tolerance policy has caused many authorities to make changes. Schools in Fairfax County have scaled back by shortening suspensions times, having them served in-school, and by using "restorative justice conferences."[34] This is where offenders take responsibility for their actions during discussions with those affected. Similar moves were made in Washington County, Maryland,[35] and in schools in New York.[36] These changes have received much criticism, especially from teachers who have been witnessing more classroom disruptions:

> New York public-school students caught stealing, doing drugs or even attacking someone can avoid suspension under new "progressive" discipline rules adopted this month [March 2015]. Most likely, they will be sent to a talking circle instead, where they can discuss their feelings . . .
>
> In Syracuse, meanwhile, teachers complain [that] student behavior has worsened since . . . restorative justice practices. They say teens are more apt to fight, mouth off to teachers and roam the halls under the more lenient policy. They're even seeing increasingly violent behavior among elementary school children . . .
>
> Recently mandated "positive interventions" have only exacerbated discipline problems in the . . . Santa Ana public school district, where middle-school kids now regularly smoke pot in bathrooms—some even in class—and attack staff—spitting on teachers, pelting them with eggs, even threatening to stab them.[37]

What is the solution to this seemingly insoluble problem? I believe the answer may lie on the other side of the Atlantic.

Before leaving Britain, we used to think everything was so far ahead in North America, especially when we were young. Everyone had big cars and color TVs in America, along with all the latest gadgets. England was *so* behind the times. However, these perceptions began to change after we moved to Canada and began taking regular trips between the old and new worlds, usually every other summer.

Our two young daughters were the first to discover that clothing styles were further ahead in the UK than back home. They would therefore stock up on the latest clothing fads during our summer sojourns, to be ahead of their Canadian classmates. Britain also led the way on the music scene—the Beatles, the Rolling Stones, Cilla Black, Tom Jones, and Pink Floyd, three of whose members were studying architecture at Regent Street Polytechnic the same three years I was there studying zoology.

And something I never realized until recently was that the latest laissez-faire approach to schooling, where the teacher's role changed from being "the sage on the stage to the guide on the side" to use the well-worn phrase, originated in England not in the United States. Its roots can be traced back to a time long before my school-teaching days of the sixties. Nor did I realize that British schoolchildren now have the reputation for being "among the worst behaved in the world."[38] This banner headline, repeated across the UK in 2014, was triggered by an academic paper written by Professor Terry Haydn, a former schoolteacher who now teaches in a department of education at the University of East Anglia. The teachers he interviewed gave some disturbing testimonials:

> You have to face serious disruption on a daily basis, pupils screaming obscenities, refusing to comply with requests to stop appalling behaviour, threatening, spitting, swearing. . . . You can't teach in the normal sense of the word, and you feel wretched for the poor kids who would like to learn but can't . . . you know you are letting them down . . .
>
> There are just too many really difficult ones . . . the system is overloaded. The poor head is at his wits end trying to get yet another senior member of staff to "mind" a kid and keep him isolated from normal classes . . .
>
> So many lessons get spoiled by low level disruption, even for experienced teachers. You feel drained by the effort of keeping on top of them.[39]

One young and disillusioned teacher, Robert Peal, wrote a newspaper article on the problem. This article coincided with the publication of his brilliant first book: *Progressively Worse: The Burden of Bad Ideas in British Schools*. In the article, he tells how, after graduating with a history degree from Cambridge University, he won a scholarship to an Ivy League university for a year. On returning from the United States, he obtained a teaching post at a secondary school, housed in an immaculate new multimillion-pound building. Puzzled why the school was often described as "deprived," he came to realize that it was bereft of *ideas*, not of material things:

> "Discipline" was treated as a dirty word. Instead, staff were encouraged to use the trendy euphemism "behaviour for learning" modishly abbreviated to B4L. . . . The results were catastrophic.[40]

Behaviour for Learning is the brainchild of educationalists Simon Ellis and Janet Tod. The article I downloaded, illustrated with the mandatory flowchart, talks about "The Three Relationships"[41] that have to be considered when dealing with a child's behavior (misbehavior?). These are their relationships with themselves, with others, and with the curriculum. Mention is also made of a child's self-esteem and giving them greater choice and "ownership of their learning."[42] Reminiscent of the new "progressive" approaches to discipline discussed earlier in this chapter, this English version was equally useless, as Robert Peal documents:

> I vividly recall one of my worst lessons descending into pandemonium: milkshake was spilt over a desk, pupils listened to music through their headphones and one girl attacked another with her umbrella.
>
> Worse, bad behaviour was actually rewarded. One boy was notorious. He came to lessons only as he pleased, swore at teachers and was an accomplished playground bully. At the end-of-year prize-giving, I was surprised to hear his name announced.[43]

He went on to explain how reward stickers were given out during lessons for merit, and how this miscreant had accumulated one of the largest caches. This was because teachers had been bribing him not to disrupt their classes.

In this school, as in many others in Britain and in North America, teachers had to abide by certain idealistic rules, all equally ineffective at curbing disruptive behavior. I would hazard that most, if not all of these policies, originated in departments of education, penned by those who never have to face such teaching problems. Educationalists have wreaked as much havoc on the teaching environment in schools as they have on the curricula that teachers are obliged to follow.

As mentioned earlier, the roots of contemporary laissez-faire policies can be traced back to England. According to Robert Peal, "If you were looking for a culprit, you couldn't do better than Summerhill School in Suffolk."[44]

Summerhill, an independent co-educational boarding school for students aged from five to seventeen, was founded by Alexander Neill and moved to its present location in 1927. Neill, a graduate in English literature, believed that children were innately wise and would develop to their full potential without any guidance from adults. His students were therefore free to do as they pleased, provided this did no harm to others. They could get up when they pleased, attend lessons as they pleased, and stay away if they wanted. Decisions about the day-to-day running of the school were made during regularly held meetings, where staff and students had an equal voice. Students were free to attend at will.

Located in a country house near the small rural town of Leiston, in East

Anglia, close to the North Sea, this small private school might never have been heard of outside of England. Indeed, if the plummeting enrollment of the late fifties had continued—numbers had fallen to about twenty-five—the school might never have survived the sixties.

All of that changed after Neill was approached by Harold Hart, an American publisher.[45] Aside from running a school, Neill was a prolific writer, with over a dozen books published by this time, both fiction and nonfiction. Hart apparently wanted to publish a compilation of his books.[46] What resulted instead was a 292-page book on education, titled *Summerhill: A Radical Approach to Child Rearing*.

Published in 1960, the book was an instant success, rising to the top of bestseller lists for nonfiction in the United States.[47] Subsequently published in Britain and elsewhere, it has sold over three million copies and it is still in print. The book unleashed a flood of interest in the school and in its maverick leader. One of the more immediate effects was an increase in enrollment, with some new students arriving from the United States.[48] Many books were written about Neill, about the school, and about this radical new approach to teaching, with such titles as *A Free Range Childhood* and *Fifty Years of Freedom*.[49]

The Summerhill laissez-faire philosophy was emulated in schools across the country as the traditional authoritarian approach to education lost ground to a more hands-off methodology that sought to empower students. The mood swing is captured in the lyrics of one of Pink Floyd's biggest hits, "Another Brick in the Wall." The chorus might have been sung by the pupils of Summerhill School, but it was recorded by students from Islington Green School, which was close to the band's studio.[50] The words, penned by Pink Floyd co-founder Roger Waters, were inspired by his hatred for the overbearing teachers at his grammar school.[51] Unlike me, my fellow Polytechnic alumnus had passed his eleven-plus exam. However, he had teachers who derided rather than encouraged their students. This is reflected in the first line of the chorus about not needing any education, the reference to the dark sarcasm they had to endure, and the repeated demand that the teachers leave the kids alone.[52]

Summerhill's founder, fêted by the luminaries of education, was recognized as one of the twelve most influential educators of the twentieth century.[53] The school flourished, attracting visitors from around the world, but its academic record was far from stellar. During a regular governmental inspection, in 1999, Summerhill received such a damning report that it was faced with closure. Resisting the temptation to go into the story that followed—which included the first time a school meeting with children had ever taken place inside a Royal Court[54]—improvements were made and the school remained open.[55] As of April 2016, there were sixty-eight students.[56]

Today Summerhill, headed by the founder's daughter, Zoe Readhead, is a far cry from the unruly school her father used to run. Although students are still free to skip lessons if they wish, there are now many rules in place—set by children rather than adults. Readhead says that the school "often now finds itself in a disciplinarian role because many children today don't have boundaries set in their homes."[57] Once referred to as a "free school," with all the connotation of anarchy, Readhead prefers to describe Summerhill as a "democratic school."

My earlier justification for including student assessment and behavior in this book is that science cannot exist where these critical issues have not been resolved. Indeed, education cannot flourish in an environment devoid of discipline, mutual respect, and high expectations: *near enough* is not good enough. I will close with a story that documents the truly remarkable consequences of following these practices.

The London borough of Hackney is one of the least desirable places to live, with street gangs, drug dealers, shootings, lootings, unemployment, poverty, and despair. The population is ethnically diverse and, on August 8, 2011, a riot broke out after police shot and killed a young black man.[58] Cars were torched, shops looted, and buildings burned by violent mobs, screaming death to the police. Thugs, some as young as eight, forced a double-decker bus to stop, attacking and robbing the fleeing passengers. Students from local schools had taken part in the mayhem, all part of the disaffected and dejected rabble. But not a single student had participated from one particular school.

Mossbourne Academy is a state school, like most other schools in England, and is housed in a modern building, which is nothing unusual these days. But Mossbourne is a truly exceptional school. Students arrive each day smartly dressed in the school uniform, ties straight, shoes polished. When seated, they all stand when the teacher enters the classroom. Teachers are addressed as "sir" or "miss" and treated with great respect.

School rules are strictly enforced, and if anyone should misbehave during a lesson they can expect to stay in detention after school, until 6 p.m.

Academic expectations are high at Mossbourne; so, too, are the students' accomplishments. In the summer of the Hackney riots, which were taking place on the school's doorstep, all but five of the final-year students were accepted into university. Remarkably, ten of the students were accepted at Cambridge University, the highest ranked university in the land. This is an achievement that precious few other schools could boast. One of the students, studying music, even turned down her Cambridge offer in favor of the Royal College of Music. Three of the students were successful in the fierce competition for acceptance into medical school. How could such remarkable successes be achieved in this state-run school where

> Around 40% of Mossbourne's pupils are on free school meals, 30% are on the special needs register, 80% are from ethnic minorities, and 40% are from homes where English is not the first language.[59]

That this should be taking place in such a poor neighborhood makes the school's stunning success even more impressive. In Hackney, crime is so rife that students are escorted to bus stops and railway stations by teachers after school each day because their smartly dressed appearance makes them targets for attack—not that they have any more money than their would-be assailants. What is the secret of this school's success? In one of his numerous newspaper interviews, the school's principal, Michael Wilshaw, explained that

> School improvement is not rocket science. What we are doing here is simply to have high expectations of our children.
>
> It's a no-excuses culture. No excuses because you come from a poor background; no excuses because you haven't done well at primary school and no excuses because you're from an immigrant family. . . .
>
> I am proud of what has happened in Hackney because we've shown that if it can be done here, it can be done anywhere.[60]

Two students shared their views with the journalist in the same article:

> Jahmala Exton, 15, lives in central Hackney, but is determined her school will help her reach the heights in a career in forensic science. She says:

> "I've been here for five years. It's much stricter than my last school and there is a lot of discipline in lessons but it's also really friendly and the teachers are nice. It's easier to learn when there is good behaviour." . . .
>
> Mitchell Osei, 16 . . . is equally optimistic about his future thanks to Mossbourne. He says: "At first I thought it was really strict but I now know that it has given me the opportunity to learn and do well. I want to be a neurologist."[61]

In a newspaper article written a few days after the riots, aptly titled "The School that Beat the Rioters," it was reported that in Michael Wilshaw's mind

> Schools and the education system have an increasingly important part to play in promoting a fair and equitable multiracial society. And he sees what he calls "the terrible incidents of these last few days" as having brought about a watershed moment, when the nation stopped making excuses for poor behaviour and poor achievement, and started looking to families and schools to work together for a more harmonious society.[62]

Mossbourne shows what can be done in troubled inner-city neighborhoods.

CHAPTER 11
AFTERMATH

The presentation I gave at the science teachers' annual meeting was in one of the larger lecture rooms in the conference hotel. Many stayed behind after my talk, mostly to share their views of the senseless curriculum. Because we had to clear the room for the next presenter, we could not tarry long, so I arranged to meet up with those who wanted to continue at a rendezvous point outside, where there were tables and chairs.

Everyone felt the frustration and irritation of having to deal with a curriculum that was so obviously alien to the science they knew. Anger was not far from the surface, but there was one teacher whose reaction has stayed with me ever since. She only spoke when we were alone, and the passion in her voice said it all. Her enthusiasm for science matched my own, and she had been passing on the flame to her students for many years. "When I close my classroom door, I teach science the way it should be taught," she confided. However, as she pointed out wistfully, she still had to test her students according to the curriculum, with all the sociological trivia and other irrelevant clutter.

Something had to be done.

That same month, I had a long chat with Bob McDonald, the national science commentator for CBC Television and CBC News Network. Bob, whom I have known for many years, is the host of *Quirks and Quarks*, a popular weekly radio program that brings listeners news of the latest happenings in science by the researchers who are conducting the investigations.

When I began my outpouring about school science, he said he knew exactly why I was so concerned. During his many years in broadcasting, he has witnessed a decline in how much school students understand about science. This is most apparent when he is visiting schools across the country to judge science fairs, which he has been invited to do now for many years. He said that he meets students with the most sophisticated of experiments but, when probed, they reveal little understanding of the underlying concepts of what they are doing.

During all the interviews he records for *Quirks and Quarks*, Bob gets to talk with a large number of science professors, both in Canada and the United States, and overseas. One of the things they all seem to complain about is how students are leaving high school unable to think for themselves. All they seem capable of doing is looking up information on the Internet. Bob succinctly summed it up: "They are information rich and knowledge poor."[1]

In answer to my question of what I could do to bring the school science problem to peoples' attention, he suggested I write an article for *Maclean's*, the Canadian weekly news magazine. I duly sent a story proposal to their editorial department, but heard nothing more. My next attempt was to write an essay for the *Facts & Arguments* page of one of our major national newspapers. Nothing more was heard from that either, so I sent copies of the article to 127 local newspapers in Ontario. I was also in contact with a popular Sunday morning program on CBC radio that includes current affairs and documentaries. At one point I thought they were going to feature my concerns about school science, but this did not materialize.

Meanwhile, the teachers who had attended my presentation gave such positive feedback to their organizing committee that I was invited to speak at the next conference. The theme in 2011 was *Science, Wise Choices, Healthy Planet*, a topic close to my heart. At the end of my talk, the association's president introduced me to a lady from the Ministry of Education. As it happens, she was an education officer in the curriculum and assessment branch. During my presentation, the focus had been on the grave problems facing our planet, and I scarcely made any mention of Ontario's science curriculum. However, she began by telling me how offended she was by my criticisms of the curriculum. When I responded that she should have been at my previous talk—she would have been *really* offended—she claimed to have all manner of research papers attesting to the excellence of Ontario's curriculum.

The president of the association, reconciliation incarnate, wanted to explore how some middle ground could be found between our two seemingly polar positions. I had the impression that he wanted the three of us to get together later on and discuss the matter, but nothing materialized. For her part, the lady from the ministry invited me to visit her office and look at all the documentation she had supporting the province's outstanding science curriculum, but I declined. It was only much later that I discovered that she had led the team that revised the science curriculum in 2005.[2]

A number of local newspapers published my article, and I received some enthusiastic responses from various parts of the province. One response, which came as a complete surprise, was from an assistant to the Minister of Education. She invited me to meet with her and discuss my concerns about the science curriculum.

On my arrival at the ministry in Toronto, located near the top of a tall downtown office tower, I was greeted by a charming young lady who was the author of the invitation. She led the way into a cavernous room that had a table with two chairs. After taking the proffered seat across the table from her, she asked me to explain the problems I had with the curriculum.

The accommodating young lady gave me her complete attention. However, attempting to explain the shortcomings of a science curriculum to someone with a BA in political science and classics was not destined for success. One item that did interest her was my mention of the errors and inaccuracies in the glossary. She asked if she could obtain a copy, and I was happy to give her the one I had brought with me. I also told her I would send her two relevant DVDs: the hands-on science presentation I gave at the museum during school break, and my presentation at the science teachers' conference. The second disc, I suggested, would leave her in no doubt as to why I had so many problems with the curriculum. While emphasizing that I had no wish to become a consultant, I expressed my interest in helping the province create a *real* science curriculum, to return scientific literacy to the classroom.

I never heard back from the ministry. However, my visit did effect one small change, as I discovered the following year (2013). While consulting the curriculum online, I was surprised to find that all but three of the twenty-nine glossary entries I had queried had been removed. They had been removed rather than corrected because that would have required someone versed in science and such people are seemingly superfluous to the needs of Ontario's Ministry of Education. Incidentally, on checking the glossary while writing this chapter, I found that all of the faulty entries had been reinstated.

While attempting to raise awareness of the serious plight of science education in Canada, I began investigating the situation in the United

States. I soon discovered the same problems there too, like relating science to society, and all the theorizing over science at the expense of actually *doing* science. I recall a commiserative telephone conversation I had with a science teacher in the States. I have long since forgotten the reason for the call, but remember commenting how someone ought to put pen to paper and write a book about the problem. It had never crossed my mind to do this at that time: the thought of tackling such an onerous task was decidedly unappealing. However, by the fall of 2013 I had concluded that I needed to write the book.

The following year, the Province of Ontario went to the polls. While there was no change in government, there was a change in the Minister of Education. The new appointee had a bachelor's degree in science, and I thought this an opportune time to contact her about the scientific failings of the curriculum. In my one-page letter summarizing what was wrong and how this could be remedied, I attached a copy of my newspaper article that had attracted the attention of her predecessor. I concluded,

> With your mandate from our Premier to look at new and innovative ways to achieve academic excellence, this would be a perfect opportunity to put science back into the science curriculum. Ontario could take the lead in North America.[3]

In her reply, she pointed out that the ministry had an ongoing schedule of curriculum review, which included an analysis of "information gathered by educators through province-wide focus group sessions."[4] Her letter concluded with the reassurance that

> all curricular documents are validated and checked for factual correctness and alignment to ministry initiatives by experts in the field from faculties of education, universities and colleges.[5]

So everything was working perfectly well, just as it always had.

CHAPTER 12

THE REALLY IMPORTANT ISSUES

In closing this book with a review of some of the most serious issues facing our planet, I hope to leave you in no doubt of the importance of scientific literacy.

Global warming would be high on most peoples' list of twenty-first century problems, and we have probably all noticed one of the most obvious effects of climate change in the extreme weather events of the last few years. Those who have not personally experienced heavy rains and floods, or extreme heat and droughts, will have heard about them in the news.

The idea that carbon dioxide released into the atmosphere from burning fossil fuels could cause climate change can be traced back to the nineteenth century,[1] but it did not enter into modern consciousness until the 1970s and '80s. There were some questions at first, but by the '90s the accumulated evidence was undeniable. In 1995, the first annual meeting of the United Nations Climate Change Conference was convened to determine how to resolve the problem. The focus of these meetings is to see what progress is being made by member states to reduce their emissions of carbon dioxide.

During the 1997 meeting in Japan, the Kyoto Protocol was reached, in which a number of member countries, including the United States and Canada, agreed to cut their emissions to certain target levels. At that time the United States was the world's leading emitter of carbon dioxide, with an annual production of over five billion tons. China was close behind, with a rate of more than three billion tons.[2] America's agreement to reduce her emissions was made during the administration of President Bill Clinton. However, when George W. Bush became president, he made it clear that he was opposed to the Kyoto Protocol. While making the point that the United States would "lead the way by advancing the science of climate change,"[3] he was not prepared to damage his country's economy by curbing carbon dioxide emissions, not while China was exempt from doing so because of her status as a developing country. Canada withdrew for similar reasons a decade later, in 2011. This was under the Conservative leadership of Stephen Harper, who had little time for science, or for scientists, especially those working on environmental issues.[4]

By 2014, China's annual carbon dioxide emissions had soared to a staggering 10.6 billion tons annually,[5] twice that of the United States. This was driven by her booming economy and by the extensive use of coal, which generates the highest carbon emissions of all fossil fuels. India, whose population is a close second behind China's, and whose economy is also rapidly expanding, ranks third behind the United States for carbon dioxide emissions. Like China, India was not held to account because of her status as a developing nation. Without some radical change, carbon dioxide emissions and global warming would continue unabated.

In the fall of 2014, a year before the 2015 climate-change conference in Paris, President Obama, addressing world leaders at the United Nations in New York, said that, "There's one issue that will define the contours of this century more dramatically than any other, and that is the urgent and growing threat of a changing climate."[6] In noting that no country was immune to the consequences of global warming, he gave some examples from the home front. These ranged from Miami being flooded at high tides, to the wildfire season in the west, which now extended throughout most of the year:

> We have to cut carbon pollution in our own countries to prevent the worst effects of climate change. . . . But we can only succeed in combating climate change if we are joined in this effort by every nation—developed and developing alike. Nobody gets a pass.[7]

Committed to avoiding a deadlock between developed and developing countries at the Paris meeting the following year, the president, along with Secretary of State John Kerry, spent considerable time in negotiations with the leaders of China and India. China's president, Xi Jinping, faced with domestic health hazards from the smog caused by burning coal, was committed to taking action. India's environment minister, Prakash Javadekar, planned a massive reforestation program to help his country meet their emission targets.[8]

The Paris meeting ended with an unprecedented accord being reached among 195 nations, including China, India, and the United States. "This is truly a historic moment," acknowledged Ban Ki-Moon, the United Nations secretary general. "For the first time, we have a truly universal agreement on climate change, one of the most crucial problems on earth."[9]

President Obama's actions were a key factor in an agreement being reached in Paris. Heightened public awareness of global warming through recent extreme weather conditions may also have played a part. However, the success or failure of the accord lies in the hands of future governments.

As far as the United States is concerned, all of the Republican contenders for the 2016 presidency were opposed to President Obama's policies on climate change and have "publicly questioned or denied the science of climate change."[10] The latter position is difficult to understand given the extent and multiplicity of the evidence upon which global warming is established.

Data for land temperatures are collected from some two thousand weather stations, located around the world. Ocean temperatures are gathered from three separate sources: thermometer readings collected aboard ships, thermometer readings obtained from buoys, and imaging data obtained from satellites.[11] Temperature records from each of these independent sources show a general trend toward increasing global temperatures since the early twentieth century, with a notable increase since the early 1980s. In addition to the temperature data, global warming is supported by the following independent evidence: the rise in sea levels due to the thermal expansion of the sea and to the addition of water from melting ice sheets; retreating glaciers; decreasing snow coverage during northern winters; acidification of oceans due to the carbon dioxide dissolving in the water; and extreme weather events.[12]

How anyone can deny the facts of global warming is difficult to understand. However, during a CNN television interview on February 20, 2014, Texas senator Ted Cruz, a presidential contender, made the remarkable statement that during

> the last 15 years, there has been no recorded warming. Contrary to all the theories that they are expounding, there should have been warming over the last 15 years. It hasn't happened.[13]

According to NASA, as reported in the world media, including the *New York Times*,[14]

> January 2000 to December 2009 was the warmest decade on record. Throughout the last three decades, the GISS [Goddard Institute for Space Studies] surface temperature record shows an upward trend of about 0.2°C (0.36°F) per decade. Since 1880, the year that modern scientific instrumentation became available to monitor temperatures precisely, a clear warming trend is present, though there was a leveling off between the 1940s and 1970s.[15]

The article went on to report that the data used were collected from three sources:

> weather data from more than a thousand meteorological stations around the world; satellite observations of sea surface temperature; and Antarctic research station measurements.[16]

Does Senator Cruz doubt the credibility of NASA? Did they not land an American on the moon in 1969, or was that all an elaborate hoax as some conspiracy theorists believe?

Denial of global warming is a logical stance to be taken by a politician from an oil-rich state like Texas. Such a person would contend that exploiting the riches of fossil fuels trapped below ground benefits everyone: more jobs, more growth, more corporate wealth, and all without affecting the climate. The same also holds true for those who accept the reality of global warming without believing it to be of man's making; according to some polls such views are held by almost half of all Americans.[17] A recent Gallup poll also showed that Americans are evenly split on the pros and cons of fracking for oil and gas.[18]

For many people, fracking is the best thing that happened since the energy crisis of the 1970s. That was when the Arab members of the Oil Producing and Exporting Countries (OPEC) imposed an oil embargo on the West. Oil prices quadrupled, motorists had to join long lines for gas, and "No Gas" signs became a familiar sight across the land. The embargo, which lasted five months, spurred domestic oil production, along with the quest for renewable sources of energy.

When mining for oil the conventional way, a vertical borehole is drilled through the rock, all the way down to the underground reservoir of oil. As the drill descends, a slurry of clay and water, referred to as drilling mud,[19] is poured into the shaft and kept continuously topped-up. This serves several functions, including keeping the drill bit cool and clear from debris and maintaining a head of pressure in the ever-deepening shaft. For every thirty-two feet descended, the pressure exerted by the mud increases by fifteen pounds-per-square-inch (one atmosphere). Maintaining a sufficient pressure in the shaft is important for preventing a blowout, the resulting gusher of which not only wastes oil but also risks igniting a fire.

In fracking, the drill's target is not a subterranean pool of oil or natural gas but an essentially horizontal layer of shale that contains locked-in oil or gas. After descending vertically to reach the shale, the drill has to be diverted horizontally, to follow the course of the seam. Once the drilling operation has been completed, high-pressure water, mixed with sand and an array of chemicals, is injected into the boring to break up the shale, thereby releasing the trapped fuel.

In the United States, thousands of wells are drilled for gas and oil fracking every year. According to estimates from the Environmental Protection Agency (EPA), during the period 2011 to 2014, between 25,000 and 30,000 new wells were drilled and fracked each year.[20] The amount of water used each time a well is fracked varies widely, but the average is about 1.5 million gallons (5.7 million liters).[21] The total annual consumption of water for fracking in the United States was estimated to be 44 billion gallons for each of the years 2011 and 2012.[22] This water is a toxic brew laced with heavy metals like mercury, cadmium, and lead,[23] along with known or suspected carcinogens.[24] Radioactive isotopes are often added, to act as tracers to monitor the flow of the water. Fracking also releases naturally occurring radioactive isotopes from the shales, some of which have long half-lives, so radioactive residues will remain in the water for millennia.

Remarkably, the EPA has no authority to regulate what happens in the fracking industry, and this includes not being informed of what chemicals are being added to the water used in the process. One of the world's leading suppliers of fracking chemicals and fluids is Halliburton, the company that pioneered fracking technology during the mid-1940s. It was because of Halliburton that fracking is no longer under the jurisdiction of the US government. The way this all came about is through a devious change in legislation that has become known as "The Halliburton Loophole."[25] Widely reported in the media, the story reads like something from a novel.

In 2005, during the presidency of George W. Bush, the Energy Policy Act was passed and became law. The infamous loophole was introduced as a small amendment to the act, made at the request of President Bush's vice president, Dick Cheney, who was the former CEO of Halliburton. Finding this revelation almost too much to believe, I downloaded the act to check it for myself. Scanning the 551-page document was not an option, so I did a word-search for hydraulic fracturing—the technical term for fracking—and found one small entry.[26] And that was all. The subtle change made by the amendment is easily missed because all it does is change how the term "underground injection" is used so that fracking fluids are not included. Accordingly, fracking was no longer under the purview of the EPA. And so it came to pass that the US oil and gas industry is free to release hazardous materials into the environment, including carcinogens, radioactive isotopes, and heavy metals—all unchecked, unsupervised, and without even having to disclose what they are. Remarkably, a Senate proposal to repeal the Halliburton Loophole was defeated in 2015, when all the Republican members but one voted against it.[27]

Some of the spent water from the fracking process is apparently reused, but most is dumped, deep underground, in abandoned oil and gas wells, or in wells specifically drilled for the purpose. Waste water from conventional oil and gas drilling is also discarded this way. These injection wells can be as deep as two miles underground and the pressure at such enormous depths—over two tons per square inch—forces the spent water into the porous rock at the bottom. Dumping toxin-contaminated water deep underground poses potential risks to aquifers that could be used for drinking water. With the number of injection wells in the United States reported to be 680,000,[28] contamination is inevitable and has already been reported. Poisoning billions of gallons of water every year and dumping this deep underground where it can contaminate clean water, threatening drinking supplies for generations to come, is an environmental disaster of catastrophic proportions. Companies should be compelled to clean fracking fluid to potable quality, using new technologies now available.[29] However, given that this is the industry that changed US legislation to serve its own ends, this will almost certainly never happen.

A small portion of spent fracking fluid is sent to water-treatment plants, but the water remains contaminated, and so does the sludge that is left behind. A recent study of such a treatment process showed that after the water was treated and dumped into a river, disturbingly high levels of radioactivity were recorded in the river sediments at the point of discharge.[30]

According to the US Energy Information Administration, the country has become the "largest producer of petroleum and natural gas in the world,"[31] with predictions that America will become a net exporter of natural gas in the near future. Fracking is here to stay, along with all the environmental problems. Aside from contaminating water, fracking has an additional environmental consequence when it is being used for extracting gas because of gas leaks into the atmosphere. Natural gas is primarily methane and, although emitting less carbon dioxide when burned than other fossil fuels (about half as much as coal),[32] when released into the atmosphere it absorbs considerably more heat from the sun than carbon dioxide. Methane is therefore a far more hazardous greenhouse gas than carbon dioxide. According to the EPA, the largest source of methane emissions into the atmosphere is the natural gas and oil industry,[33] mostly from leaks at gas wells. Given that the number of actively producing gas wells in the United States was over half a million in 2014,[34] the scale of the problem is immense. Many wells produce gas as well as oil, and those that are mined solely for oil also release some methane during the fracking process.

The oil embargo stimulated interest in renewable energy, and wind power received much of the attention. Denmark took the lead in research and development and has manufactured and installed wind turbines in seventy-four countries around the world.[35] According to Denmark's official website, "more than 40 percent of Denmark's energy supply comes from wind power,"[36] the intention being to reach 50 percent by the year 2020. On a *per capita* basis, Denmark leads the world in wind-powered electricity generation, while China, the major emitter of greenhouse gases, generates the largest *total* amount of electricity, with the United States a close second.[37] Regardless of these seemingly impressive numbers, wind turbines account for less than five percent of electricity generation in the United States. Coal-fired and natural-gas-fired generating stations, in contrast, each account for 33 percent, with nuclear power contributing 20 percent.[38] There is, however, much variation between states, with Iowa and South Dakota each generating more than 25 percent of their electricity from wind power.[39]

Modern wind turbines typically have three blades, each 60 to 120 feet (20 to 40 meters) long, and they generate about 2.5 to 3 megawatts of power.[40] This is enough electricity to supply some 700 homes in the United States, or about 1500 in Europe.[41] (I believe the wide discrepancy reflects how wasteful of electricity we are on this side of the Atlantic.) Each rotor blade is an airfoil, like the wings of an airplane, and is remarkably long and slender, reminiscent of a sailplane's wings. As we saw in chapter 4, long slender (high aspect-ratio) wings generate maximum lift for minimum drag.

Figure 12.1: Wind turbines have long slender blades, like the wings of gliders, to generate maximum lift—the turning force—with minimum drag.

The turbine is supported on a tall tower, not only for ground clearance but also to raise it above the level of the slower moving winds below. Turbines do not generate power until wind speeds reach about 8 mph (13 kph), a gentle breeze, when flags begin to flutter. Maximum power is reached in winds of about 22 mph (35 kph), and turbines have to be shut down when speeds exceed about 56 mph (90 kph), to prevent their being damaged.[42] Winds only blow at optimal speeds for a minor part of the twenty-four-hour period—usually at night, when power demands are low—which is a major disadvantage of wind power. Locating wind turbines offshore, where winds are often more frequent and more powerful, can improve the situation. The largest turbines are used in offshore locations, and one of biggest is built by the Danish company Vesta. It is difficult to comprehend a wind turbine standing atop a tower 460 feet (140 meters) high, with blades that are 260 feet (89 meters) long.[43] This 8-megawatt giant could potentially supply 4,500 Danish homes.

Billions of dollars are being invested in wind power, and the prospects for this renewable source of energy in reducing greenhouse gases looks promising. However, in *Green Illusions*, a disquieting new book about clean energy, Ozzie Zehner paints a rather different picture. In an analysis of facts reminiscent of what I experienced when untangling real from perceived science in the *Framework* document (see chapter 5), he made a disturbing discovery. He shows how wind-energy proponents and consultants, all with vested interests in the growing industry, had their overly optimistic estimates of the power output of wind turbines incorporated into a US Department of Energy (DOE) report on wind power. Being associated with a reputable body like the DOE lent legitimacy to their claims. And, because so few people understood the underlying technicalities, the public, along with politicians and journalists, got carried away with the idea that wind power could make a greater contribution to supplying clean energy than it can in reality.[44]

The power generated by wind turbines increases with air density so, for a given wind speed, more electricity would be generated on a cold winter's day than in summer. Given that power increases with density and that water is 800 times denser than air, placing turbines in running water is a logical progression. Oceans cover 70 percent of the Earth's surface and are in perpetual motion, driven by winds and by tides. Extracting tidal energy therefore makes far more sense than exploiting wind energy.

The world's first marine turbine, built in Britain, was tested in the summer of 2003 in shallow water off the Devonshire coast, in the southwest of England. Named *SeaFlow*, the two-bladed rotor is 36 feet (11 meters) in diameter. The generator to which it was attached was mounted on a mobile frame that could move up and down the tower, which was partially embedded in the seabed. The tower extended above the surface so the turbine could be removed from the sea for maintenance. Aside from undergoing testing and being raised for refitting with improved components, it remained submerged for the next six years, unobtrusively converting the sea's tidal surges into electricity.

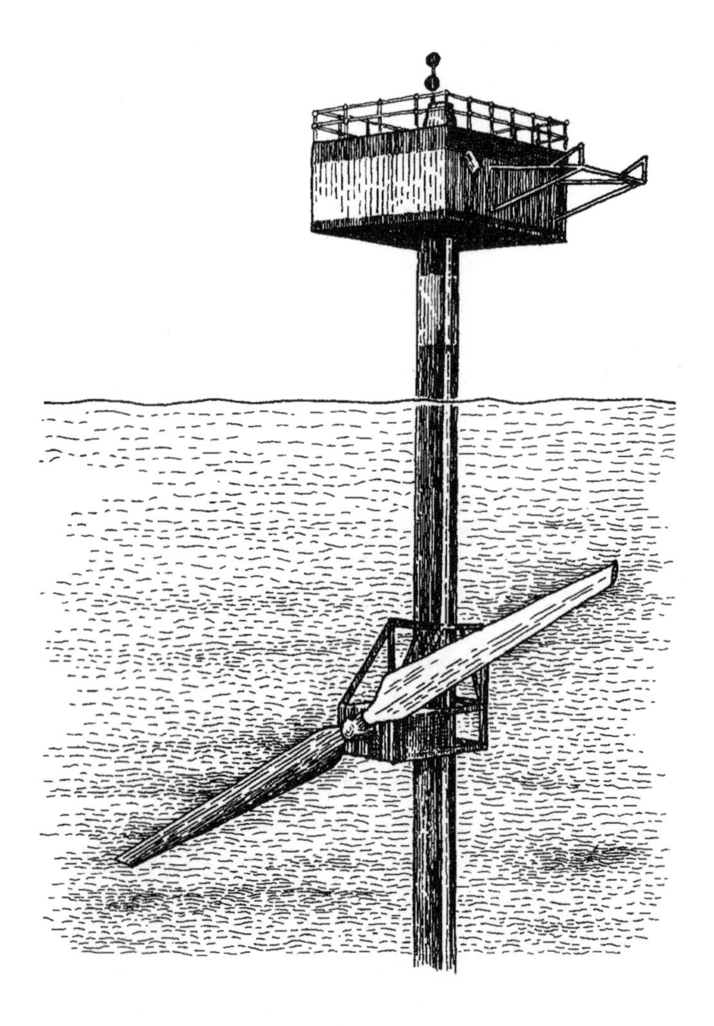

Figure 12.2: *SeaFlow*, the world's first marine turbine, was installed off the Devonshire coast of England in 2003. There it remained for the next six years, generating electricity from tidal power.

There are two tides during the twenty-four-hour period, and the blades were adjustable, so that power could be generated during both the flow tide and ebb tide. Therefore, although being smaller and generating only one-fifth the power of a wind turbine, *SeaFlow* did this for much of the day. Furthermore, as tide times vary with local geography, if marine turbines were placed in different locations around the coast, power could be generated throughout the entire twenty-four-hour period. Tide times also advance by about an hour each day, giving even more variation in the timing of peak generation times.

The next-generation version, named *SeaGen*, has two twin-bladed rotors, mounted side-by-side on a horizontal arm that can be moved up and down on the tower, like its predecessor. The base of the tower is attached to a three-legged platform, the legs of which extend into holes drilled in the rocky seabed. The rotors are larger than those of *SeaFlow*, being 52 feet (16 meters) in diameter.[45] The power output is rated as equivalent to that of a 2.4 megawatt wind turbine.[46]

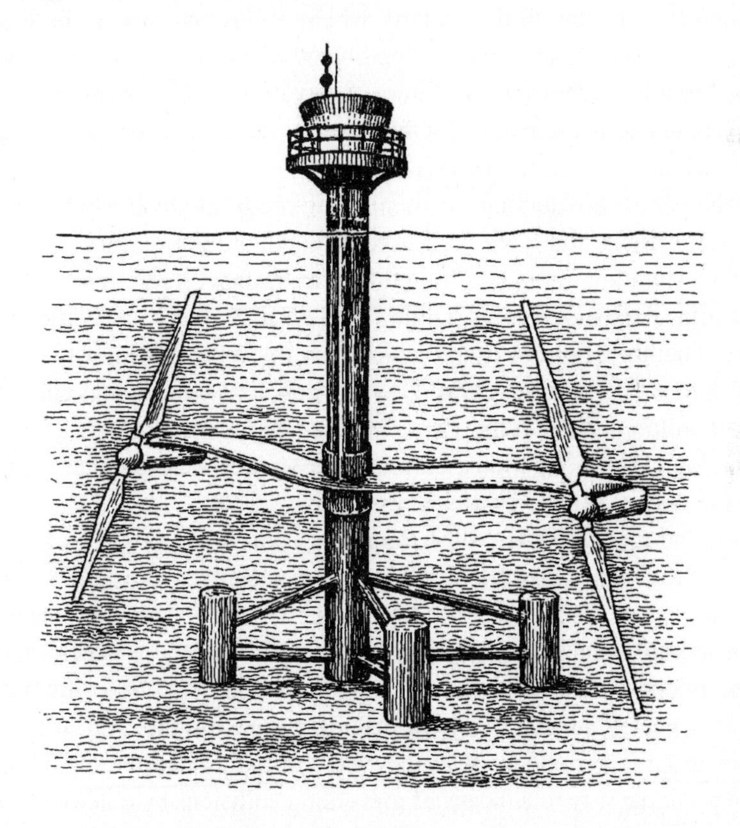

Figure 12.3: *SeaGen*, built by the same company as *SeaFlow*, has two twin-bladed rotors 52 feet (16 meters) in diameter.

The new turbine was installed off the coast of Northern Ireland, in the Narrows of Strangford Lough, in May 2008, and fed electricity into the

National Grid, making it the first commercial-scale marine turbine in the world.[47] To satisfy the law, an environmental impact assessment had to be conducted. One of the main criticisms that have been made of marine turbines is the potential damage to the marine environment, particularly the harm that could be done to marine animals like seals and porpoises, through coming into contact with the moving blades, especially with the tips, which are moving the fastest. The final report of the assessment ran to seventy-seven pages.[48] Unsure of what I might find, I began reading.

My first surprise was that monitoring began a few weeks before the turbine was installed. This was to make a survey of the birds, mammals, and other organisms in the vicinity where the turbine was to be located, before any changes took place. The objective was to provide an environmental baseline against which future observations could be compared after the turbine had been installed. Observations continued for more than five years, which was another surprise.

The people conducting the monitoring program obviously knew what they were doing and were well used to conducting biological fieldwork. One investigation involved fitting thirty-six seals with electronic GPS tags to monitor their movements. Diving surveys and video observations were made of benthic (bottom-living) organisms, everything from barnacles and starfish to soft corals and crabs. Diving birds, like gannets and guillemots, were monitored to see if they were affected by the installation. Aside from the biological investigations, current-flow studies were conducted, to see how the turbine affected water movements.

Marine mammals, and other large vertebrates that could be at risk of striking the rotors, were monitored using sonar. Between July 2008 and July 2011, a total of 1948 potential targets were detected. The crew responded by temporarily shutting down the turbine in 342 of these occurrences, as a precautionary measure.[49] Some would argue that this artificially reduced the incidence of animal collisions, making marine turbines appear less hazardous to large vertebrates than they were in reality. On the other hand, it could point the way to a means of preventing collisions by automatic detection and temporary shutdowns. The findings of the sonar survey showed that marine mammals could be detected within 50 meters (164 feet) of the turbine, allowing the rotor to be stopped before they reached it.[50]

The overall environmental impact of installing the turbine was minimal. Seals and porpoises continued to pass through the narrows, with no decline in their numbers. Seals still hauled up on shore the same as before, and there were no significant differences in their movements. The only observable changes were small modifications in the behavior of seals and porpoises, suggesting a degree of avoidance of the device. There were no marine mammal deaths attributable to physical contact with the turbine. The foraging behavior of diving birds showed no evidence of their being displaced from important feeding areas by the turbine, but they did appear to avoid diving close to the installation. There were no significant changes in benthic communities, nor were there any significant changes in water flow in the narrows due to the turbine. Although the tower created a wake on the surface, this did not extend far down, which may have explained the absence of significant changes in benthic ecology.

For turbine critics still concerned over potential collision hazards with rotating blades, there is an alternate turbine design, engineered by a second UK company, named OpenHydro. These marine turbines are multibladed, reminiscent of an aircraft engine, and are enclosed so that the tips of the blades are not exposed. Significantly, there is a large central opening through which large marine animals, and even a careless human swimmer,

Figure 12.4: This radically different design of marine turbine has a multibladed rotor that is completely enclosed, reminiscent of the engine of an airliner. The central opening is large enough to allow most marine animals to pass through without any harm

could easily pass, thereby avoiding any contact with rotating blades. These turbines are designed to rest on the seabed, where they would have little if any environmental consequences. Trials have been conducted in the Bay of Fundy, off Canada's east coast, to exploit the remarkably powerful currents. Here, the 16.3-meter (54 feet) tidal range is the highest in the world.[51] Unfortunately, the turbine was damaged by the high speeds of the flow, but more trials are apparently being planned.[52]

Readers might be surprised to learn that tidal power has been exploited in Britain for almost a millennium. Eling Tide Mill, which still stands beside the sea on England's Hampshire coast, was mentioned in the Domesday Book (1086). Operating like a traditional water mill, where water is diverted into a millpond, the incoming tide is used to flood a reservoir. When the tide is at its highest and the reservoir is full, sluice gates are closed to prevent the seawater from escaping. Then, as the tide falls, the water is channeled to drive the water wheel. Once the tide turns and begins to rise, the sluice gates are opened and the cycle is repeated. Built for grinding grain, Eling Tide Mill is still being used to mill flour today, though not with the original equipment. Similar tide mills were built elsewhere in Britain, in Europe, and along the eastern seaboard of North America. With the advent of steam power these became redundant, but times have changed again and there is now a renewed interest in tide mills.

Using marine turbines to extract energy from the ceaseless motion of the sea offers the greatest opportunities for reducing carbon emissions, with minimal environmental consequences. However, the same is not true for the other options.

Solar power, long promoted as the green solution to the energy problem, has always concerned me because of the toxic material used in the solar panels, like cadmium, arsenic, and lead. The panels do not last forever and, in spite of any directives for proper disposal, most will likely end up in landfill sites where the toxins will leach into the environment. Toxins are only part of the problem with solar power, as chronicled in *Green Illusions*, to which readers are directed for more information.

Some years ago, I heard a radio program where the argument was made that hydrogen was the clean fuel of the future. Hydrogen would be powering all our vehicles, and our economy. In this new hydrogen-based economy, the hydrogen gas would be used in fuel cells, along with oxygen, to generate electricity. The electricity would then be used to power the electric motors driving our vehicles. The only emission released into the atmosphere would be water vapor. How does this work?

For those who remember learning about electrolysis at school, what happens inside a fuel cell is the exact reverse. Electrolysis is the splitting of water, H_2O, into its two components, hydrogen and oxygen, by passing electricity through it. Inside a fuel cell, hydrogen and oxygen combine to form water, releasing electricity in the process.

If you have never carried out electrolysis yourself, it is very simple. All you need is a 9-volt battery (as used in smoke detectors), two lengths of wire, about five inches (12 cm) long (you can use lengths of twist-tie, stripping off an inch or so of the plastic at either end), and a small glass of water. After attaching the wires to the battery terminals, lower the other ends into the water, side by side and about half an inch (1 cm) apart. Within seconds you will see bubbles streaming off one of the electrodes. This is hydrogen. Oxygen is released at the other electrode but, since there is only half as much oxygen in water (H_2O) as hydrogen, the bubbles are less apparent.

An electric vehicle that is powered by a fuel cell has to carry a tank of compressed hydrogen, and another of compressed oxygen. The two gases are delivered to the cell, where they combine together to form water, generating electricity in the process. In the new hydrogen-based economy, hydrogen could also be used directly, as a fuel, and burned inside a modified internal combustion engine, instead of using gasoline.

Powering vehicles with hydrogen made as little sense to me back then, listening to that radio program, as it does today. However, as if to prove me wrong and show that hydrogen was the way to go, fleets of hydrogen-powered buses began appearing on the streets in Europe, the UK, and in Australia, between 2003 and 2009. The year 2009 was also when Boeing demonstrated a single-seat electric airplane that ran on fuel cells. And when Vancouver hosted the Winter Olympic Games in 2010, British Columbia rolled out a fleet of twenty fuel-cell buses to transport people back and forth to the events. Using hydrogen might seem a sensible solution, but advocates overlook some fundamental facts.

The first problem with the hydrogen economy is that about 95 percent of hydrogen is manufactured by heating methane (natural gas) with steam, using fossil fuels to provide the heat. To make things worse, the process generates carbon dioxide as a by-product. Furthermore, the conversion of hydrogen and oxygen into electricity inside a fuel cell is only about

60 percent efficient.[53] The conversion of electrical power into mechanical power inside the electric motor that drives the vehicle is more efficient, but is still less than 100 percent. The net result of all this is that the environmentally friendly fuel-cell bus, whispering down the road emitting only water vapor, has left behind a much larger carbon footprint than a regular bus, running on diesel fuel and belching out exhaust fumes. While all this is really bad for the environment, it is very good for the gas industry supplying the methane to the hydrogen plant.

Electric cars, where a rechargeable battery supplies power to the motor, are likely purchased to reduce global warming. However, the effectiveness of this all depends on how the electricity was generated. For much of the United States, electricity is primarily generated using fossil fuels, so most electric-car owners in America still leave a sizeable carbon footprint. Fortunately, this is not the case in every state. In South Dakota, for example, wind and hydroelectric generation accounted for 73 percent of the total electricity generation in 2015, so electric car owners there can feel they are helping.[54] In Canada, hydroelectricity accounts for about 60 percent of total production, but there is wide variation across the country.

The detrimental effects of climate change are rapidly increasing and most of us have witnessed, or seen news coverage, of some aspects of global warming. These are exemplified by extreme weather events, including flooding, high winds, droughts, and forest fires. As I was writing this chapter, the forest fire that forced the evacuation of 88,000 people from Fort McMurray, a remote town in northern Alberta, was still growing out of control. Driven by hot dry weather and exacerbated by the wind, the fire, which had been burning for a week, was about 620 square miles (almost 400,000 acres) in area: about twice the size of New York City and one and a half times larger than Hong Kong. The loss of all those trees, along with the extra carbon dioxide released into the atmosphere, contributed even more to global warming.[55] It is a cruel irony that Fort McMurray is the center of Canada's tar-sand industry, one of the dirtiest sources of petroleum in the world. Aside from destroying or damaging hundreds of square miles of boreal (northern) forest, polluting the air from the burning of fossil fuels used during the extraction process, and contaminating billions of gallons of water, the end product is extremely high in emissions.

The effects of global warming have been devastating in human terms, with disasters like Fort McMurray, the 2015 Texas floods, and the extreme heat of the same year that killed thousands in India and Pakistan. However, there is one far more serious aspect of global warming that has received scant attention, and this is the effect that global warming has on plankton.

Although most people have probably heard of plankton, I suspect few know much about it. And I would be surprised if many realize how vital plankton is to life on our planet. This is because phytoplankton—the microscopic plants in the plankton—while accounting for less than one percent of the total plant biomass, produces half of the planet's oxygen.[56] Phytoplankton also absorbs half of the planet's carbon dioxide—more than 100 million tons every day. Plants (and certain bacteria) provide the *only* mechanism on Earth for absorbing carbon dioxide and producing oxygen, something to remember when felling trees to build more houses. Animals and plants are mutually dependent, as we saw in chapter 8. Without plants, life as we know it would cease to exist.

With the advent of satellite imagery, it is now possible to measure the relative abundance of plankton by measuring the greenness of oceans, the color being due to the chlorophyll in the microscopic plants. This technique has been providing data since 1979. Data from traditional methods of assessing plankton abundance date back to the early 1900s. By combining the two data sets, researchers have been able to measure the changes that have been taking place in plankton abundance for more than the last one-hundred years. A disturbing picture has emerged.[57] Plankton has declined in eight of the ten ocean regions studied, a pattern that correlates with increasing sea-surface temperatures. As the phytoplankton declines, so too does the uptake of carbon dioxide, accelerating global warming even further. If plankton decline continues unabated, the consequences are unthinkable.

Meanwhile, *Homo sapiens*, the wise man, plunders the oceans. He has factory ships, massive vessels equipped with freezers and processing plants to convert the hundreds of tons of fish caught every day into supermarket packages. Many of these ships use midwater trawling nets. Guided by sonar, the huge net can be set at the same depth as a school of fishes spotted on the screen. The whole school can then be scooped up, along with anything else that happens to be lying in its path. As one fisherman said to me, aboard a traditional fishing boat, many years ago, "the fish aren't getting any smarter."

Some ships fish by longlining, where baited hooks are suspended, at intervals, from a line, up to sixty miles (100 km) long, paid out from the ship's stern. The line, with several thousand baited hooks, is kept afloat by a series of buoys. Once the line has been set, it is cut loose and allowed to drift for several hours, often overnight. Tuna and swordfish are caught by longlining, and plastic glow sticks are usually attached to the line, at intervals, to attract them to the light. These plastic tubes are filled with chemicals that become luminescent when the stick is activated by bending it. Designed for one-time use, discarded ones frequently end up in the sea, along with their toxic contents. Another ecological consequence of longlining is that sea turtles, which are known to be attracted to light, are often hooked as an unwanted by-catch. So too are many seabirds, including albatrosses. Sharks are also caught and, if tuna is the target catch, these are discarded, usually already dead from drowning.[58]

Many years ago, I was on a research vessel longlining off the Atlantic coast of Canada. The lines were set in the evening and left adrift overnight while we slept. On one occasion when the line was hauled in we discovered one swordfish, which was dead. All the rest were sharks, and most of these were alive. One of the many survivors was a blue shark (*Prionace glauca*), a magnificent animal with its sleek streamlined form and radiant blue color. Remarkably flexible, the twelve-foot (3.7m) shark was bending so far to one side, trying to escape the hook, that the tail touched the snout. To see the swimming motion of living sharks is an impressive sight. (Blue sharks, like many others belonging to this particular family, can be dangerous. Members of this group are referred to as requiem sharks, a wonderfully evocative name.) The swordfish, in contrast to the sharks, was as solid as a tree trunk, so much so that it bore my weight when I tried sitting on it.

The sea has become a dumping ground for much more beside glow sticks. Huge swaths of floating debris, mostly plastic, litter oceans around the world, dumped overboard or washed out to sea from land. Much of this becomes broken up into small particles, some of which is ingested by copepods (chapter 8), and other minute animals in the zooplankton, thereby entering the food web.[59] Plastics contain toxins, like bisphenols, which contaminate the water as well as the living organisms.

Discarded plastic is found everywhere in the world these days. On an ornithological fieldtrip to Patagonia, over thirty years ago, we were finding Styrofoam and other bits of plastic washed up on beaches all the way down to Tierra del Fuego. Parenthetically, I have an abhorrence for Styrofoam; why can't the meat and fish departments in supermarkets use pulp cartons, like they do for eggs, instead of this awful stuff? For an intriguing if disturbing account of the plastic scourge, you should read *Plastic Ocean*, a book written by an inquisitive sailor.[60] The most distressing images in the book for me were photographs of the remains of dead albatross chicks, each showing the remnants of the gut filled with plastic trash, mostly bottle caps.

At the rate at which oceans are being plundered, fish stocks will crash, along with the industries pursuing them. We saw the same thing happen in Canada with the east-coast cod fishery. People have been fishing for cod in the waters around Newfoundland since the 1600s. Remarkably, some even crossed the Atlantic in galleons, arriving from Britain and Europe each year to catch cod, salt them, and take them back home at the end of the season. But overfishing, which began in the 1960s, led to declining stocks, and in 1992 a moratorium was declared; the fishery was closed and more than 35,000 people lost their jobs.[61]

Quite apart from overfishing, there is considerable waste in the fishing industry. Capelin caught off Newfoundland, for example, is sorted by sex, and the females, ripe with roe (eggs), are packaged, frozen, and exported, mostly to Japan. The males, every bit as edible as the females, are discarded and used as fertilizer. The worst in waste is the Chinese practice of catching sharks, slicing off their fins for soup, and throwing the finless fishes overboard to die. In addition to threatening some shark species with extinction, this abhorrent behavior is having some unexpected ecological side effects. These include the collapse of several stocks of bivalves, like scallops.[62] The reason for this is that skates and rays, which feed on shellfish, have increased in numbers due to the demise of their predators, the sharks. This underscores the far-reaching consequences of altering the fine ecological balance in nature.

One last and most disturbing example of how the sea is being plundered is the harvesting of plankton, for use in food supplements of all things. When I first saw adverts for Jameson's krill oil extract[63]—krill is a small, shrimp-like crustacean, about two inches (50 mm) long—I thought this harvesting was a fairly recent phenomenon. However, according to the Food and Agriculture Organization of the United Nations,[64] the industry dates back to the 1960s. Using large midwater trawling nets, the annual catch far exceeds one-hundred-thousand tons. Imagine how many trillions of individuals this represents and what ecological effects this must be having, especially given the precarious state of the plankton.

The damage wreaked on land is far more apparent than what has been done in the oceans. One example that has attracted considerable attention is the deforestation of rainforests. While the rate of destruction has slowed, it still continues, and for what? Most of the clearing in the Amazonian forest has been to make room for cattle farming, along with tree-felling for sought-after hardwoods like mahogany. There is also gold mining, mostly along the river banks, which are blasted away with water cannons to expose potential seams of gold. Trees are chain-sawed along the way, turning lush riverine habitats into barren industrial wastelands.[65] Mercury is used in the extraction process and, while some is apparently recovered and reused, much of the toxic element flows downstream to contaminate the environment.

Open-pit mining, conducted in countries around the world, has not attracted as much attention as deforestation, but the damage being done is considerable. Excavating for everything from uranium and diamonds to copper and coal, mining has scarred the face of the planet with lunar landscapes large enough to be seen from space. Aside from the aesthetics, and the pollution, and the destruction of plants that would otherwise be offsetting carbon emissions, is the loss of habitats and the inevitable loss of species.

The extinction of species has been going on since life appeared on the planet, 3.5 billion years ago. Extinction is as natural, and inevitable, as the appearance of new species, which offset the losses of those that disappear. However, there have been times during Earth's long history when the rate of extinction has been so high that the event is recognized as a mass extinction. This is said to occur when more than three-quarters of the species have been lost within a geologically short interval.[66] Five mass extinctions have been recognized in the fossil record, the last being at the end of the Cretaceous Period, 65 million years ago, when the last of the dinosaurs disappeared.

If the current rate of extinction continues unabated, the end result is inevitable, as discussed in a study by a group of biologists and paleontologists in a paper titled, "Has the Earth's Sixth Mass Extinction Already Arrived?"[67] What sets the present wave of extinctions apart from all the others is the remarkable speed, and the fact that it is attributable to a single species, our own.

To illustrate the rapid pace of change, consider the alarming rate at which rhinoceroses are being exterminated, and all because some people believe in the medicinal qualities of their horns. In 2007, thirteen rhinos were poached in South Africa. By 2015 that number had accelerated to 1,175.[68] Elephants are under similar threat for their tusks, used for carving ornaments. The estimated number of elephants killed in Africa from 2010 to 2012 is a staggering 100,000.[69] Since 2009, the elephant population in Tanzania alone has declined by more than 60 percent.[70] The extermination of these two symbols of Africa is perpetrated by the usual suspect: greed. Rhino horn is selling on the black market for more than twice the price of gold. The current demand is being driven by Vietnam rather than China, apparently because of a rumor that swept the country that a politician's cancer had been cured by powdered rhinoceros horn.[71]

Why is our species so destructive, and why are we destroying life on this planet at such an alarming and unparalleled rate? Simply put, there are too many people in the world. To help put this into perspective, we need to consider what happens to species in the natural world.

American robins, which reach maturity in their first year, lay about ten eggs annually and can live for a dozen years. One female could therefore produce a hundred offspring in her lifetime, creating a dynasty of thousands. We could expect to be tree-deep in robins, but nature does not work that way. This is because animal populations are regulated by natural forces, like predation, disease, and food limitations. Not so for *Homo sapiens*. We circumvent nature with technological guile—industrializing food production, harvesting some species, domesticating others, exterminating those deemed pests, and eradicating disease.

When Shakespeare died, four hundred years ago, the world population was about 560 million,[72] less than half that of present-day India. For the next two centuries, population growth remained modest, and fairly linear, as it had done since medieval times. However, at about the time of the Industrial Revolution, which began in the late 1700s, growth became exponential, rocketing to the present population of seven billion.

Significantly, the increase is more rapid in developing countries than in the rest of the world. This is because the *total fertility rate*—the number of live births per woman—is higher there. The highest fertility rates occur in Africa, the top five nations being Niger (7.63), Somalia (6.61), Mali (6.35), Chad (6.31), and Angola (6.20).[73]

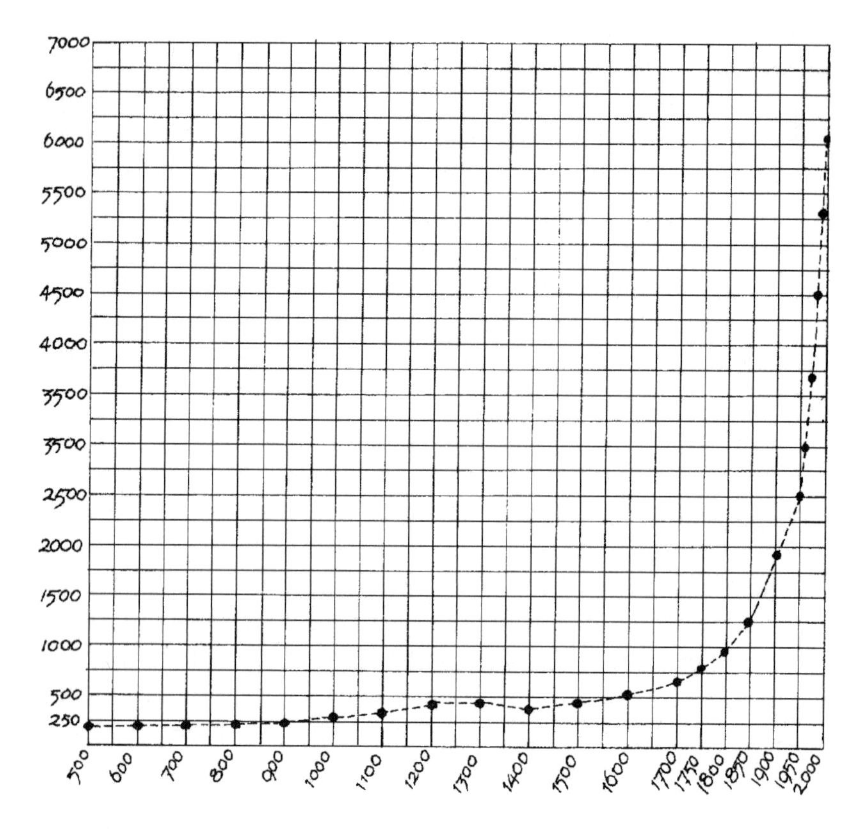

Figure 12.5: The world population, shown in millions on the left side of this graph, increased relatively slowly throughout the Middle Ages and beyond. Then, at about the time of the Industrial Revolution, it increased exponentially. The small downward dip after 1300 is attributable to the Black Death (bubonic plague), which killed about half of the European population.

Somalia, with its droughts and food shortages, has been in the media spotlight for many years. The inhospitable terrain of scrub and desert is clearly unsuited for supporting large numbers of any species, but the struggle to survive and raise large families continues. Heart-wrenching images of starving children and helpless parents are difficult to ignore, and the developed world has been swift to respond. But what should that response be? Celebrities like Bono, lead singer of the rock group U2, have been front and center in pressuring governments to give more aid, but does that really help places like Somalia?

In her recent book, *Dead Aid*, Zambian author Dambisa Moyo argues that foreign aid to Africa—over one trillion US dollars since the 1940s—has done more harm than good.[74] Dr. Moyo makes a compelling argument that aid discourages independence, suppresses local initiatives, and encourages rampant corruption, as exemplified by Mobutu, the former president of Zaire. During his thirty-two-year dictatorship, Mobutu embezzled an estimated five billion dollars from aid funds to support his lavish lifestyle.[75] Meanwhile his people starved.

Bono, in a 2011 TV interview with CNN's Anderson Cooper, talked about increasing aid to Somalia and improving crops to yield more food, but made no mention of reducing fertility rates.[76] Nor was contraception on the agenda for African aid during the administration of George W. Bush. During Bush's presidency, billions were spent on AIDS relief in Africa, but none of the funds were allowed to be spent on contraceptives. This makes absolutely no sense, especially given the way that AIDS is transmitted.[77]

Countries like Somalia, where six or more children per family is the norm, are somewhat reminiscent of the developed world during the nineteenth century, except that modern medicine is now available. Darwin, for example, had ten children, two of whom died at birth. People had large families back then because birth control was neither widely available nor practiced, and because fewer children survived, due to poor sanitation and primitive medicine. Advances in the knowledge of diseases and how they are spread led to an awareness of the importance of proper sanitary practices, both in medicine and in everyday life. Physicians began washing their hands before attending patients or delivering babies, and antiseptics were used for cleaning wounds and surgical instruments. Much needed improvements were also made in sewage systems, which helped protect drinking water from contamination. Child mortality fell and, with the availability of contraceptives, couples had fewer children.

A perusal of total fertility rates across the world yields some unexpected results. Brazil, for example, with the world's highest Catholic population, ranks 158, with a fertility rate of only 1.8, while India, the world's second most populous country, ranks 78 with a rate of 2.5.[78] During the 1950s, India's fertility rate was close to 6.0, and a concerned government, under the leadership Prime Minister Nehru, launched the world's first family-planning program, in 1952. Little progress was made in the early years, especially in rural areas where most of the population lived. Then, in the 1970s, with fertility rates still above 5.0, the government began a controversial program of mass sterilization.[79] Conventional contraceptive methods were also practiced and fertility rates fell.

China, the country with the largest population, now enjoys one of the lowest fertility rates, at 1.6, giving it a world ranking of 181.[80] Surprisingly, China's fertility rate began plummeting during the 1960s, more than a decade before implementation of the one-child policy in 1979.[81] This was achieved by encouraging couples to have fewer children, rather than compelling them by law. Regardless of the efforts of China and many other countries, there are still over one hundred countries producing more children per family than that required to maintain the status quo, and the world population is predicted to reach over nine billion by 2050.[82]

How many people does the world need? Shall we continue until the entire globe is as crowded as Mumbai, with a species diversity comparable to downtown Manhattan? We delude ourselves with the notion of *sustainable development*—a concept as illusory as the hydrogen economy—but there are far too many people in the world already. We have seen some of the consequences of our overabundance, and the time to take serious steps to reverse this most urgent threat of global warming has long since passed. However, the most important items on most political agendas are the economy, economic growth, and the creation of more jobs.

The notion of growth is entrenched in the commercial consciousness. But the resulting environmental degradation can no longer be ignored. Inaction is not an option, but our best efforts to reverse the damage will have no lasting effect until we reject the canon of growth. The economic turmoil of the last few years is surely the writing on the wall that continued economic growth is no longer tenable.

The perils of population growth were heralded over two centuries ago in a book by Robert Malthus: *An Essay on the Principle of Population.* "The power of population is indefinitely greater than the power in the earth to produce subsistence for man," he wrote, prophetically.[83] "Population, when unchecked, increases in a geometrical ratio. Subsistence increases only in an arithmetical ratio." To illustrate his point, he did a computation to show how the current population of seven million Britons would outgrow the island's food production long before a century had passed.[84] The consequence of uncontrolled growth was "that premature death must in some shape or other visit the human race."[85] Premature death could take many forms, from pestilence and plague to the "inevitable famine [that] stalks in the rear."[86] His essay on the eternal struggle for survival influenced a number of nineteenth-century intellectuals, including Darwin.

In 1968 Paul Ehrlich wrote a chilling bestseller, *The Population Bomb*, sounding the alarm over the population explosion. Like Malthus, his primary focus on the consequences of uncontrolled population growth was famine. This was encapsulated in the first sentence of the prologue to his book:

> The battle to feed all of humanity is over. In the 1970s the world will undergo famines—hundreds of millions of people are going to starve to death in spite of any crash programs embarked upon now.[87]

Ehrlich, dismissed by some as a doomsayer, was wrong about famines of global magnitude, but his book, which sold two million copies, drew attention to the problem of overpopulation.[88] He also pointed to the negative effects this was having on the planet. A recent video by the *New York Times*, replete with clips from the past, shows the extent of Ehrlich's influence at the time his book was published.[89] UK science writer Fred Pearce, noting how birthrates have been "falling dramatically,"[90] voices the opinion of many others that the *Population Bomb* has been defused. While fertility rates have dropped worldwide—from 4.5 births in 1970 to 1975, to 2.5 for 2010 to 2015[91]—the world's population will continue growing, reaching a predicted size in excess of 11 billion by 2100.[92]

The argument is often made that by using modern agricultural practices, together with cultivating land that has not yet been *developed* (to use the popular euphemism for destroying green space to make money), it would be possible to feed these additions. Even if we could feed four billion more people, which I find difficult to believe, what would be the cost to the rest of the planet, and to the quality of human life?

Much of the conflict and misery of the present times is caused by overpopulation, and the resulting lack of resources and prospects that this brings. When people are crowded together with no jobs and no hopes or aspirations for the future, mayhem is inevitable. Some months ago, I heard part of a radio interview with a journalist who had interviewed a captured jihadist in the Middle East. The salient point I remember was that he was one of seventeen siblings.

The other side of the coin to impoverished regions with booming birthrates are affluent parts of the world, like Europe, the UK, Japan, and the United States, where fertility rates are below 2.1, the rate required to maintain current population levels. This causes concern for some people, like Stewart Brand, the American writer who appeared in the recent video with Paul Ehrlich. In the video, Brand had the last word:

> The point at which population peaks around 9 billion in the 2040s or 50s the story will not be, "Oh my God, we've got 9 billion people, how horrible." It will be, "Oh my God, we're running out of people."[93]

This is not the way I see things. For me, a less crowded planet than the current one with seven billion people would be a far better place. A world where elephants can freely roam, and where albatrosses can scoop up squid instead of plastic waste.

Facing and resolving the global problems of the twenty-first century requires individuals with a solid foundation in science, knowledge that is based on learning from first-hand practical experience, not from abstract theorizing. Youngsters should spend as much time as possible engaged in practical work. Learning about photosynthesis, for example, should involve experimenting with living plants, not using "a model to illustrate how photosynthesis transforms light energy into . . . chemical energy."[94]

If students spent more time *doing* science rather than *theorizing* over it or just reading about it, they would be motivated to take an interest in the subject. Instead, students are being turned off science in school. My intuitive grandson, who completed his seventh grade while I was finishing this chapter, told me that his class conducted only one practical during the entire school year. And all they did was weigh a cup of water, stir in a quantity of salt, and then reweigh it to see if there had been any gain in weight. When I asked what else they did during their science classes he said that their teacher gave them handouts of text to read; she also read out sections from a science book. Without any prompting from me, he said his science classes were boring, a view shared by his classmates. I suspect the same is true in schools across North America. Little wonder that fewer students are going on to study science at university.[95]

Universities have also been moving away from teaching science through hands-on experience, and mention has been made that many medical schools have discontinued dissecting cadavers (chapter 5).[96] Undergraduates in biology are similarly spending less time in the laboratory, and little or no time in the field.[97] The annual marine biology course that I taught was discontinued soon after I retired, even though students rated this as the best course of their entire university careers. What better way could there be for learning about the ecological balance between organisms, and the importance of protecting and preserving the natural environment, than by going out into the field and experiencing the living world firsthand?

Our two-week field course, primarily for students in their final year, was divided into two equal segments. During the first week, we took the students to various locations to collect specimens from different marine habitats, from rocky and muddy shores, to an estuary where freshwater meets the sea. We also took them on a boat cruise, where we trawled the seabed for benthic organisms, and collected plankton samples from the surface layers of the sea. For many students this was their first field course, and we showed them how to *look* and how to *observe*. I would lift up handfuls of seaweed attached to rocks to show them what they might otherwise miss—whelks, winkles, worms, sponges, crabs, barnacles, limpets, sea squirts, mussels, and all the rest. Once they had collected samples of the different kinds of seaweeds and animals, it was back to the lab, where they learned how to identify them, so they could be recognized and named on subsequent occasions. We all worked hard, and the students were so animated by the diversity of life and by the sheer wonder in what they were seeing, that many were reluctant to leave the lab when we turned off the lights and sent them off to their beds.

Having become familiar with different organisms and their habitats, the second week was spent working on their individual research projects, which accounted for the bulk of the marks for the course. Most of the students did really well, impressing us with how much they had learned and how capable they were of conducting independent research—observing, testing, and questioning as scientists. But there was one notable exception, a student who raised the alarm bell for me that all was not well with the way some students learned science at school.

We made a point of talking to each of the students about their planned project before they got started, to make sure that what they had in mind was feasible. This particular student decided to investigate the effects of exposing sea anemones, collected from tide pools along the shore, to sea water of different salinities. However, they had a problem that they could not resolve. To keep marine organisms alive in the lab, their containers had to be supplied with running seawater, which was available on tap. Given that the student wanted to use diluted seawater, how could this be done? Somewhat surprised that the student could not have worked this out, I supplied the obvious solution. After placing the anemones in a small container, just large enough to house them, this would be filled with seawater diluted to the desired salinity. A fish tank, raised a few feet above the specimens, would then be filled with the diluted seawater, which would slowly be siphoned into first container to maintain a gentle flow. Having explained all this, I was surprised when the student asked me what a siphon was.

Sharing the content of the conversation with my colleague, I said it seemed unlikely the student would perform well and receive a good grade in our course. Wondering aloud whether they would graduate, I was surprised to learn that the student had not only received a bachelor's degree but was now working toward a master's qualification. That a student so lacking in practical skills could have progressed so far in science underscores the problem in our education system, a problem that begins in our schools.

Until real science returns to the classroom, captivating childhood curiosity, youngsters will remain disinterested, and few will pursue the subject at university. Scientific literacy will remain as elusive as ever, leaving a citizenry unable to see the environmental problems for what they are, unable to separate reality from rhetoric, unable to distinguish science from nonsense. Countries and corporations will continue plundering and polluting, destroying habitats and extinguishing other species in the process. But it does not have to be this way.

Science and scientific literacy is too important for us to leave school science in its present dysfunctional state. Changes must be made. What is taught in the classroom has to be determined by those who are accomplished in the subject, not by scientifically illiterate educationalists.

NOTES

INTRODUCTION

1. Editorial Board, "Who Says Math Has to Be Boring?" *New York Times*, December 7, 2013, http://www.nytimes.com/2013/12/08/opinion/sunday/who-says-math-has-to-be-boring.html (accessed June 23, 2016).

2. *A Framework for K–12 Science Education* can be downloaded at https://www.nap.edu/catalog/13165/a-framework-for-k-12-science-education-practices-crosscutting-concepts?gclid=COjK1-Cgus8CFQMIaQodlkkEHw. The Next Generation Science Standards can be downloaded at https://www.nap.edu/catalog/18290/next-generation-science-standards-for-states-by-states?gclid=CLaf_6Ohus8CFQytaQodtG0L0A (both last accessed October 2, 2016)

3. "Eleven-Plus Exam," *Wikipedia*, https://en.wikipedia.org/wiki/Eleven-Plus_exam (accessed June 28, 2016).

4. New genera: *Excalibosaurus*, *Leptonectes*, *Hudsonelpidia*.

New species: *Excalibosaurus costini*, *Leptonectes moorei*, *Leptonectes solei*, *Macgownia janiceps*, *Shastasaurus neoscapularis*, *Stenopterygius cuneiceps*, *Stenopterygius macrophasma*, *Temnodontosaurus eurycephalus*. I hasten to point out that the generic name *Macgownia* was named by one of my former graduate students, Ryosuke Motani, not by myself. Dr. Motani is now a full professor at the University of California, Davis.

5. The certificate, dated March 9, 2004, reads, in part, "The Natural Sciences and Engineering Research Council of Canada recognizes Chris McGowan for important research achievements that have contributed to the sum total of human knowledge and the advancement of the economic and social well-being of Canadians for over 25 years of NSERC's existence."

CHAPTER 1: FROM CRUSHED CANS TO CURRICULUM CONSULTANTS

1. For more information on airfoils, see Christopher McGowan, *A Practical Guide to Vertebrate Mechanics* (Cambridge, UK: Cambridge University Press, 1999): 220–21.

2. "Crushed Cans and Broken Bones," YouTube video, 6:38, posted by "excalibosaurus" on March 31, 2010, https://www.youtube.com/watch?v=y90gbpjv1J4 (accessed

July 27, 2016). For more information on the bone-breaker, see McGowan, *Practical Guide to Vertebrate Mechanics*, p. 100.

3. The Ontario curriculum is available online as three separate documents, according to grades:

Ontario Ministry of Education, *The Ontario Curriculum Grades 1–8 Science and Technology, 2007*, http://www.edu.gov.on.ca/eng/curriculum/elementary/scientec18currb.pdf (accessed May 30, 2016).

Ontario Ministry of Education, *The Ontario Curriculum Grades 9 and 10 Science, 2008*, http://www.edu.gov.on.ca/eng/curriculum/secondary/science910_2008.pdf (accessed May 30, 2016).

Ontario Ministry of Education, *The Ontario Curriculum Grades 11 and 12 Science, 2008*, http://www.edu.gov.on.ca/eng/curriculum/secondary/2009science11_12.pdf (accessed May 30, 2016).

4. Ontario Ministry of Education, *Ontario Curriculum Grades 1–8*, p. 5.

5. Ibid., p. 36.

6. G. S. Aikenhead, "STS Education: A Rose by Any Other Name," in *A Vision for Science Education: Responding to Peter Fensham's Work*, ed. Roger Cross (New York: RoutledgeFalmer, 2003), pp. 59–75.

7. The two quotes defining STSE education are attributable respectively to Avi Hofstein and others, and to Dr. Robert Yager, Professor of Science Education at the University of Iowa. Both quotes are cited on p. 482 of: Nasser Mansour, "Science-Technology-Society (STS): A New Paradigm in Science Education," *Bulletin of Science, Technology and Society* 5, no. 1 (2009): 482–97, https://ore.exeter.ac.uk/repository/bitstream/handle/10871/11347/a%20new%20paradigm.pdf?sequence=2 (accessed May 30, 2016).

8. As an alternative to estimating the volume of the helium balloon, it can be measured directly by collecting the helium released as it is deflated. To do this, I first bent a short length (about one foot, or 30 cm) of quarter-inch copper pipe (from a hardware store) into a U-tube. One end was attached to a short length of rubber tubing. This was later used to direct the gas into the mouth of an upturned measuring jug that had been filled with water in a water-filled sink. While pinching off the neck of the balloon to prevent the gas from escaping, the other end of the tube was inserted a small distance (about an inch, or 2 cm), into the mouth of the balloon. Gas leaks were prevented using an elastic band that had been looped, several times, over the end of the U-tube before the balloon was attached. While still pinching off the gas, the bundled elastic band was rolled over the neck of the balloon, securing it to the pipe. With the help of an assistant, the gas can be gently released into the upturned jug (or measuring cylinder if available), displacing the water. Once the container has been filled with helium by the downward displacement of water, the jug can be refilled with water and the procedure repeated, as the volume of helium inside the inflated balloon is equivalent to several jugs.

9. Ontario Ministry of Education, *Ontario Curriculum Grades 1–8*, p. 5.

10. Ontario Ministry of Education, *Ontario Curriculum Grades 9 and 10*, p. 5.

11. H. Lynn Erickson, *Concept-Based Curriculum and Instruction: Teaching Beyond the Facts* (Thousand Oaks, CA: Corwin Press, 2002), p. 7.

12. "Biography," Lynn Erickson C & I Consulting, http://www.lynnerickson.net/biography/ (accessed May 30, 2016).

13. H. Lynn Erickson, "Concept-Based Teaching and Learning," International Baccalaureate Position Paper, International Baccalaureate Organization, 2012, http://www.ibmidatlantic.org/Concept_Based_Teaching_Learning.pdf (accessed May 30, 2016).

14. Erickson, *Concept-Based Curriculum and Instruction*, p. 164.

15. Grant Wiggins and Jay McTighe, *Understanding by Design* (Alexandria, VA: ASCD, 1998), p. 10.

16. Grant Wiggins and Jay McTighe, *The Understanding by Design Guide to Creating High-Quality Units* (Alexandria, VA: ASCD, 2011), p. 71.

17. Ibid., p. 120.

18. Ontario Ministry of Education, *Ontario Curriculum Grades 1–8*, p. 6.

19. Ibid.

20. Ibid., p. 115.

21. Ibid.

22. Ibid.

23. Ibid., p. 116.

24. Ibid.

25. Ibid.

26. Ibid., p. 152.

27. McGowan, *Practical Guide to Vertebrate Mechanics*, p. 199.

28. Ibid., p. 220.

29. Ontario Ministry of Education, *Ontario Curriculum Grades 1–8*, p. 164.

30. Ibid., p. 159.

31. Ibid., p. 155.

32. Ibid., p. 116.

CHAPTER 2: THE YOUNG SCIENTIST

1. For further reading about memory and stressful events see, L. Cahill and J. L. McGaugh, "Mechanisms of Emotional Arousal and Lasting Declarative Memory," *Trends in Neuroscience* 21 (1998): 294–99.

Also see Christa K. McIntyre and Benno Roozendaal, "Adrenal Stress Hormones and Enhanced Memory for Emotionally Arousing Experiences," in *Neural Plasticity and Memory: From Genes to Brain Imaging*, ed. Federico Bermudez-Rattoni (Boca Raton: CRC Press/Taylor & Francis, 2007), http://www.ncbi.nlm.nih.gov/books/NBK3907/ (accessed June 3, 2016).

2. B. S. Grant, "Fine Tuning the Peppered Moth Paradigm," *Evolution* 53, no. 3 (1999): 980–84.

For further reading on the peppered moth, see: B. S. Grant, "Intentional Deception: Intelligent Design Creationism," *eSkeptic: The Email Newsletter of the Skeptics Society*, 2004, http://www.skeptic.com/eskeptic/04-06-01/ (accessed June 3, 2016).

Jonathon Wells, "The Peppered Myth: '"Of Moths and Men'" an Evolutionary Tale," *Discovery Institute*, September 30 (2002), http://www.discovery.org/a/1263 (accessed June 3, 2016). The Discovery Institute is a creationist site for promoting the notion of intelligent design.

J. Mallet, "The Peppered Moth: A Black and White Story after All," *Genetics Society News* 50 (2004): 34–38.

L. M. Cook, and J. R. G. Turner, Decline of Melanism in Two British Moths: Spatial, Temporal an Inter-Specific Variation, *Heredity* 101 (2008): 483–89.

3. Sting, *Broken Music* (New York: Dial Press, 2003).

4. "De Havilland Comet," *Wikipedia*, https://en.wikipedia.org/wiki/De_Havilland_Comet (accessed June 3, 2016).

5. Ibid.

6. "Civil Aircraft Accident: Report of the Court of Enquiry into the Accidents to Comet G-ALYP on 10th January, 1954 and Comet G-ALYY on 8th April, 1954," *Ministry of Transport and Civil Aviation*, 1955, http://lessonslearned.faa.gov/Comet1/G-ALYP_Report.pdf (accessed June 3, 2016), p. 11.

7. Ibid., p. 14.

8. Ibid., pp. 15–16.

9. Ibid., p. 17.

10. Ibid.

11. Ibid.

12. Ibid., p. 18.

13. Ibid., p. 14.

14. For a discussion of fatigue in glass see, Keith B. Doyle and Mark A. Kahan, Design Strength of Optical Glass, *Proceedings SPIE Optomechanics*, 5176 (2003):1–12, http://www.sigmadyne.com/sigweb/downloads/SPIE-5176-3.pdf (accessed June 3, 2016).

There is also a good general article on fatigue fracture at: "Fatigue (Material)," *Wikipedia*, https://en.wikipedia.org/wiki/Fatigue_(material) (accessed June 3, 2016).

15. "De Havilland Comet," *Wikipedia*.

CHAPTER 3: IGNORANCE IS BLISS

1. William Rand, "How Does Reiki Work?" from *Reiki: The Healing Touch* (Southfield, MI: Vision Publications, 1991), International Center for Reiki Training, http://www.reiki.org/FAQ/HowDoesReikiWork.html (accessed June 3, 2016).

2. Anna Sharratt, "Naturopaths' Prescribing Rights Expanded: After Extensive Lobbying Efforts, Naturopaths across Canada are Getting Governmental Green Lights for Greater Prescribing Rights," CBS News, October 30, 2009, http://www.cbc.ca/news/technology/naturopaths-prescribing-rights-expanded-1.821015 (accessed June 3, 2016).

3. "FAQ," Association of Accredited Naturopathic Medical Colleges, https://aanmc.org/naturopathic-medicine/faq/ (accessed June 3, 2016).

4. "Acupuncture," Newmarket Naturopathic & Integrative Health Clinic, http://www.newmarketnaturopath.com/#!acupuncture/cher (accessed June 3, 2016).

5. "Fall 2016 Courses," University of Minnesota Center for Spirituality & Healing, http://www.csh.umn.edu/education/credit-courses (accessed June 3, 2016). This course appears to be no longer offered, but the description can be viewed at https://www.csh.umn.edu/education/credit-courses/csph-5643-horse-teacher-intro-nature-based-therapeutics-equine-assisted-activities-therapies-eaat (accessed October 4, 2016).

6. Robert Tomlinson and Silvana Fazzolari, *BIE: History and Research*, Institute of Natural Health Technologies, 2007, http://torontoholisticinspiration.com/wp-content/uploads/2014/04/BIE-History-and-Research-revised-Feb3-2012.pdf, p. 10 (accessed June 3, 2016). Tomlinson and Fazzolari cofounded the Institute of Natural Health Technologies and BIE.

7. Dianne Sousa, "Alphaghetti Credentials: 'Registered Holistic Allergists' and 'BioEnergetic Intolerance Elimination,'" *Skeptic North*, September 17, 2012, http://www.skepticnorth.com/2012/09/a-tale-of-alphaghetti-credentials-%E2%80%9Cregistered-holistic-allergists%E2%80%9D-and-%E2%80%9Cbioenergetic-intolerance-elimination%E2%80%9D/comment-page-1/ (accessed June 4, 2016).

8. A full account of Perkins and his tractors is given by: W. S. Miller, "Elisha Perkins and His Metallic Tractors," *Yale Journal of Biology and Medicine* 8, no. 1 (1935): 41–57.

9. Haygarth's original publication can be seen at http://www.jameslindlibrary.org/haygarth-j-1800/ (accessed June 3, 2016). However, the old English text, where the letter s appears as f, makes for challenging reading. The quote from Haygarth, written in modern form, is taken from p. 511 of A. J. M de Craen et al., "Placebos and Placebo Effects in Medicine: Historical Overview," *Journal of the Royal Society of Medicine* 92, no. 10 (1999): 511–15, http://www.ncbi.nlm.nih.gov/pmc/articles/PMC1297390/pdf/jrsocmed00004-0023.pdf (accessed June 3, 2016).

10. For further reading on the placebo effect in Parkinson's disease see, R. de la Fuente-Fernández et al., "Expectation and Dopamine Release: Mechanism of the Placebo Effect in Parkinson's Disease," *Science* 293 (2001): 1164–65.

R. de la Fuente-Fernández, S. Lidstone, and A. J. Stoessl, "Placebo Effect and Dopamine Release," *Journal of Neural Transmission [Supplementa]* 70 (2006): 415–18.

11. For further reading on the genetic component of the placebo effect see: K. T. Hall, J. Loscalzo, and T. J. Kaptchuk, "Genetics and the Placebo Effect: The Placebome," *Trends in Molecular Medicine* 21, no. 5 (2015): 285–94.

12. T. J. Kaptchuk et al., "Components of Placebo Effect: Randomized Controlled Trail in Patients with Irritable Bowel Syndrome," *BMJ* 366, no. 7651 (May 3, 2008): 999, http://www.bmj.com/content/bmj/336/7651/999.full.pdf (accessed June 4, 2016).

13. T. J. Kaptchuk et al., "Sham Device *v* Inert Pill: Randomized Controlled Trial of Two Placebo Treatments," *BMJ* 332, no. 7538 (February 18, 2006): 391, http://www.bmj.com/content/332/7538/391?ehom= (accessed June 4, 2016).

14. Stephen Barrett, MD, "Be Wary of Spinal Decompression Therapy with VAX-D or Similar Devices," *Chirobase: Your Skeptical Guide to Chiropractic History, Theories, and Practices* (revised August 21, 2015), http://www.chirobase.org/06DD/vaxd/vaxd.html (accessed June 4, 2016).

15. John Gibson, "Parents of Toddler Who Died of Meningitis Used Home Remedies Rather than Consult Doctor, Court Hears," CBC News, March 7, 2016, http://www.cbc.ca/news/canada/calgary/jury-trial-truehope-toddler-dies-trial-underway-1.3479460 (accessed June 4, 2016).

16. N. Nelkon and P. Parker, *Advanced Level Physics* (London: Heinemann Educational Books, 1977), p. 45.

17. See, for example, Charles J. Ammon, "An Introduction to Plate Tectonics," Earthquake Seismology, Penn State, July 31, 2001, http://eqseis.geosc.psu.edu/~cammon/HTML/Classes/IntroQuakes/Notes/plate_tect01.html.

CHAPTER 4: PREPARED FOR THE WORST

1. After removing a wing from a dead bird, it has to be spread out and pinned to a sheet of cardboard or Styrofoam to dry out. After a few days, the pins can be removed; the wing will retain its stretched-out shape. Liberally painting the stub and both sides of the entire wing with rubbing alcohol will eliminate any health concerns regarding its handling.

2. Gary Bradshaw, "Otto Lilienthal: (1848–1896)," Mississippi State University Inventor's Gallery, http://invention.psychology.msstate.edu/i/Lilienthal/Lilienthal.html (accessed June 4, 2016).

3. More information about the camber of airfoils can be found at several websites, including: ALLSTAR (Aeronautics Learning Laboratory for Science, Technology, and Research) Network, http://www.allstar.fiu.edu/aero/flight31.htm (accessed June 4, 2016). Also see: Christopher McGowan, *A Practical Guide to Vertebrate Mechanics* (Cambridge: Cambridge University Press, 1999), pp. 220–29.

4. The cross-sectional area is properly referred to as the *frontal area.*

5. The term *soaring* is often used interchangeably with *gliding*, but the two terms are not the same. Gliding is a passive mode in which height is continuously lost, whereas in soaring flight height is gained or maintained by extracting energy from the movements of air.

6. The mass of a solid object varies with its volume, and volumes change with the cube of the length. The volume of a cube with an edge of 2 cm is $2 \times 2 \times 2 = 8$. The volume of cube that is twice as large is $4 \times 4 \times 4 = 64$. The increase in volume, hence in mass, is $64/8 = 8$. This relationship may be written as: m ∝ (size)3.

7. The aspect ratio of the Wright Flyer is given in: John Anderson, "Wings: From the Wright Brothers to the Present," National Air and Space Museum, December 17, 2011, http://blog.nasm.si.edu/aviation/wings-from-the-wright-brothers-to-the-present/ (accessed October 26, 2016).

8. One of the small paper wings was square with an edge of 1½ inches (39 mm), giving an aspect ratio of one. The other was 3 inches (75 mm) long and ¾ inch (19 mm) wide, giving an aspect ratio of four. Each had the same profile, with a maximum depth of ⅜

inch (5 mm). This was achieved by giving the square wing a fold-down flap that was 5 mm wide, while that of the narrow wing was only 2 mm wide.

9. Just recently I discovered that small electric blowers for keyboards are no longer available. For those wishing to replicate the experiment with the two small paper airfoils, I found that blowing air with the lips, placed about 6 inches (15 cm) away from each airfoil, produced the same results as when using the small blower. A gentle steady blow keeps the high aspect-ratio airfoil airborne, but the other airfoil fails to lift its trailing edge clear from the surface.

10. Many papers have been published on the remarkable flying performance of albatrosses. See, for example: P. Jouventin and H. Weimerskirch, "Satellite Tracking of Wandering Albatrosses," *Nature* 343 (1990): 746–48.

G. Sachs, J. Traugott, A. P. Nesterova, and F. Bonadonna, "Experimental Verification of Dynamic Soaring in Albatrosses," *Journal of Experimental Biology* 216 (2013): 4222–32, http://jeb.biologists.org/content/216/22/4222 (accessed June 4, 2016).

11. Inclined planes are frequently referred to as inclined plates; plate and plane are interchangeable.

12. Ridges and dimples can be reduced or removed from the aluminium pie plate by placing it on a hard surface and rubbing firmly with the back of a spoon.

13. If the angle of attack of an inclined plane is gradually increased, the lift increases, but so too does the drag. The lift reaches a maximum at an angle of attack of between 10° and 20°, and then decreases. Drag increases all the time, reaching a maximum at an angle of 90°, at which point the lift is zero. (See McGowan, *Practical Guide to Vertebrate Mechanics*, p. 224.)

14. The Plasticine block I used was 2½ × 1¼ × ¼ inch, which was adequate. I also prepared a permanent base by pouring plaster of paris into a small cardboard box of suitable size. By placing a slab of Plasticine on the bottom of the box before adding the plaster, the three shish-kebab sticks can be correctly set up before pouring commences.

15. Ontario Ministry of Education, *The Ontario Curriculum Grades 1–8 Science and Technology, 2007*, p. 116, http://www.edu.gov.on.ca/eng/curriculum/elementary/scientec18currb.pdf (accessed October 6, 2016).

16. For more information about Mary Anning, see Christopher McGowan, *The Dragon Seekers* (Cambridge: Perseus Publishing, 2002).

17. When making artificial sedimentary rock, I found Uhu Twist & Glue worked well. However, it does foam up during the shaking, creating a large "head" when poured into the cup. This can be reduced by gentle blowing. Elmer's white glue probably works better than their clear Earth Friendly School Glue, which leaves a thick surface layer after pouring.

18. Elmer's Earth Friendly School Glue leaves a glutinous residue on the surface.

19. The trackway exhibited in the former dinosaur gallery of the ROM was a cast of the original material.

20. If difficulty is found in obtaining a clear footprint, it is because the sand is too dry. Add a few drops of water, thoroughly mix, and try again.

CHAPTER 5: THE BEST LAID PLANS

1. The first two quotes, about the NGSS, are from the first page of the introduction, in the NGSS front matter section: "The Next Generation Science Standards: Executive Summary," National Academy of Sciences, 2012, http://www.nextgenscience.org/sites/ngss/files/Final%20Release%20NGSS%20Front%20Matter%20-%206.17.13%20Update_0.pdf (accessed November 24, 2015).

2. The quote from the president was given in April 2013, during the Third Annual White House Science Fair, which can be seen at: "Educate to Innovate," Education for K–12 Students, White House, https://www.whitehouse.gov/issues/education/k-12/educate-innovate (accessed November 11, 2015).

3. National Research Council, *A Framework for K–12 Science Education: Practices, Crosscutting Concepts, and Core Ideas* (Washington, DC: National Academies Press, 2012), p. 1, http://www.nap.edu/catalog/13165/a-framework-for-k-12-science-education-practices-crosscutting-concepts (accessed November 11, 2015).

4. Ibid., p. ix.

5. Ibid., p. 2.

6. Ibid., p. 1.

7. Ibid., p. 10.

8. Ibid., p. 11.

9. Ibid., p. 2.

10. Ibid.

11. Ibid., p. 30.

12. Ibid., p. 3.

13. Ibid., p. 84.

14. Ibid., p. 3.

15. Ibid., p. 28.

16. Ibid.

17. Ibid., p. 31.

18. Ibid., pp. 24–25.

19. Ibid., p. 33.

20. Although the card works, it curls when in contact with the snow. I found a broken plastic plate in the recycling box to be ideal for the job. So too was a flat-bottomed plastic tray, used for containing meat.

21. National Research Council, *Framework for K–12 Science Education*, p. 43. The numbers within the square brackets are the original note numbers found in the text for the sources cited in the *Framework*.

22. Richard Lehrer and Leona Schauble, "Cultivating Model-Based Reasoning in Science Education," in *The Cambridge Handbook of the Learning Sciences*, ed. R. K. Sawyer (Cambridge, England: Cambridge University Press, 2006), p. 371.

23. Ibid., p. 383.

24. R. K. Sawyer, ed., *The Cambridge Handbook of the Learning Sciences* (Cambridge, England: Cambridge University Press, 2006), p. xi.

25. Ibid.

26. Ibid.

27. National Research Council, *Framework for K–12 Science Education*, p. 44.

28. M. Ford, "Disciplinary Authority and Accountability in Scientific Practice and Learning," *Science Education* 92, no. 3 (2008): 405.

29. Ibid., p. 406.

30. Leema K. Berland and Brian Reiser, "Making Sense of Argumentation and Explanation," *Science Education* 93, no. 1 (2008): 26.

31. National Research Council, *Framework for K–12 Science Education*, p. 44.

32. D. Klahr and K. Dunba, "Dual Space Search During Scientific Reasoning," *Cognitive Science* 12, no. 1 (1988): 32.

33. C. V. Schwarz, B. J. Reiser, E. A. Davis, L. Kenyon, A. Achér, D. Fortus, D, Y. Shwartz, B. Hug, and J. Krajcik, "Developing a Learning Progression for Scientific Modeling: Making Scientific Modeling Accessible and Meaningful for Learners," *Journal of Research in Science Teaching* 46, no. 6 (2009): 632–54.

34. Klahr and Dunbar, "Dual Space Search," p. 32.

35. C. McGowan, "The Ichthyosaurian Tailbend: A Verification Problem Facilitated by Computed Tomography," *Paleobiology* 15 (1989): 429–36.

36. National Research Council, *Framework for K–12 Science Education*, p. 43.

37. Schwartz et al., "Developing a Learning Progression for Scientific Modeling."

38. Lehrer and Schauble, "Cultivating Model-Based Reasoning in Science Education."

39. Ibid.

40. National Research Council, *Framework for K–12 Science Education*, p. 59.

41. Ibid., p. 43.

42. Lehrer and Schauble, "Cultivating Model-Based Reasoning in Science Education," p. 371.

43. Berland and Reiser, "Making Sense of Argumentation and Explanation," p. 26.

44. National Research Council, *Framework for K–12 Science Education*, p. 43.

45. Ibid., p. 60.

46. Ibid., p. 61.

47. Ibid., p. 77.

48. Ibid., p. 63.

49. Ibid., p. 74.

50. Ibid.

51. J. R. Martin and R. Veel, eds., *Reading Science: Critical and Functional Perspectives on Discourses of Science* (London: Routledge, 1998).

52. J. R. Martin, "Recontextualisation, Genesis, Intertextuality and Hegemony," in *Reading Science*, ed. Martin and Veel, p. 5.

53. National Research Council, *Framework for K–12 Science Education*, p. 58.

54. Ibid., p. 75.

55. Ibid., p. 76.

56. Ibid., p. 77.

57. Ibid., p. 58.

58. Lehrer and Schauble, "Cultivating Model-Based Reasoning in Science Education," p. 37.

59. National Research Council, *Framework for K–12 Science Education*, p. 63.

60. Ibid., p. 57.

61. Ibid., p. 59.

62. Ibid., p. 68.

63. Ibid., p. 61.

64. Ibid., p. 50.

65. Christopher McGowan, *A Practical Guide to Vertebrate Mechanics* (Cambridge: Cambridge University Press, 1999), p. 1.

66. National Research Council, *Framework for K–12 Science Education*," p. 60.

67. Ibid., pp. 237–38.

68. Ibid., p. 84.

69. Ibid., p. 92.

70. Ibid., p. 93.

71. Ibid., p. 99.

72. Ibid., p. 100.

73. Provided the weights added to the pan are not too heavy, the pan would return to its original position after they were removed. The wire would therefore not be permanently stretched, and its behaviour is described as being elastic. For more information, see McGowan, *Practical Guide to Vertebrate Machines*, pp. 21–26.

J. E. Gordon, *Structures: Or Why Things Don't Fall Down* (Boston: Da Capo Press, 2003), pp. 38–40 (first published in 1978).

74. See McGowan, *Practical Guide to Vertebrate Machines*, p. 27.

75. For more information on strain energy, see McGowan, *Practical Guide to Vertebrate Machines*, pp. 68–70.

76. National Research Council, *Framework for K–12 Science Education*, p. 1.

77. Ibid., p. 220.

78. Some might quibble that the book *Science at the Supermarket* is a science publication, elevating the science count to nine. However, from my reading of the outline of the book, whose senior author is an educational psychologist, I would disagree.

79. Some of the educational publications have multiple citations in the reference listings given at the end of each chapter and in the appendices. While this inflates the disparity between the numbers of science and educational publications, it is the citation counts that reflect the influence of these publications on the *Framework*. Incidentally, the non-science sources also include two engineering papers.

80. National Research Council, *Framework for K–12 Science Education*, p. x.

81. Benchmarks for Science Literacy is available online at: http://www.project2061.org/publications/bsl/online/index.php?txtRef=&txtURIOld=%2Ftools%2Fbenchol%2Fbolframe%2Ehtm (accessed January 5, 2016).

82. Ibid., chapter 13, p. 1.

83. Ibid.

84. Ibid., p. 2.

85. Ibid.
86. Ibid.
87. Ibid.
88. Ibid.
89. Ibid., p. 3.
90. Ibid.
91. Ibid. The illustration of the map can be seen by clinking on the link "View map."
92. Ibid., p. 4.

CHAPTER 6: TO SIR, WITH LOVE: TEACHING SCHOOL IN THE SIXTIES

1. Regent Street Polytechnic was granted university status in 1992, becoming the University of Westminster. As a university, it can award its own degrees. Previously, students had to register as external students of the University of London and sit the University's external exams for their degrees.

Because of my unorthodox educational background—failing to go to a grammar school and having to attend a technical college after leaving school to obtain the rest of my A-level GCEs—my applications to study at university were unsuccessful.

2. Honors degrees in England are rated at four levels: first, upper second, lower second, and third class. Candidates attaining the pass mark without a high enough grade to warrant honors are designated as having simply passed. At that time, final exams were just that, and success or failure hinged entirely on a set of exams taken at the end of the third year. We had six three-hour written exams: one in the morning and one in the afternoon, for three consecutive days. The following week we had four three-hour practicals, set during two consecutive days.

3. See endnote 1.

4. I had used a 10 percent solution of acetic acid, the same acid I was using for fossil preparation (described later in the chapter). This dissolved the eggshell much faster than using vinegar, which is the same acid but is more dilute. I have subsequently tried using a 10 percent solution of nitric acid, one of the strong mineral acids, but this denatured the membrane, causing it to become almost impermeable to water.

It takes about twenty hours to dissolve the shell with vinegar. At this point, the last remnants can be can be removed by *gently* rubbing with a plastic scourer, rinsing off under slowly running water. Changing the vinegar halfway through speeds the process.

A strong sugar solution is made by adding sugar to boiling water in a saucepan until it just begins to thicken. The saucepan is then allowed to stand in cold water until the solution is at room temperature.

5. Christopher McGowan, *The Dragon Seekers* (London: Little, Brown, 2002), pp. 13–31.

6. *Blue Peter*, a BBC television program that first aired in 1958, is still running today.

It is said to be the longest-running children's TV show in the world: https://en.wikipedia .org/wiki/Blue_Peter (accessed July 19, 2016).

CHAPTER 7: TREADING FAMILIAR GROUND

1. When arranged according to topic, the Standards occupy 102 pages; when arranged by "Disciplinary Core Idea," they occupy 103 pages, for total lengths of 445 and 446 pages respectively. This material can be seen at: http://www.nextgenscience.org/next -generation-science-standards (accessed June 6, 2016). One of the options given by clicking on "Search the standards" is to download a PDF arranged by topic.

2. A listing of the writing team is given at: http://www.nextgenscience.org/ writing-team (accessed June 8, 2016). Clicking onto a name gives a biographical sketch of the member, identifying those who were team leaders.

3. "Appendix H—Understanding the Scientific Enterprise: The Nature of Science in the Next Generation Science Standards," Next Generation Science Standards, p. 2, http:// www.nextgenscience.org/sites/ngss/files/Appendix%20H%20-%20The%20Nature%20 of%20Science%20in%20the%20Next%20Generation%20Science%20Standards%20 4.15.13.pdf (accessed January 23, 2016).

4. Ibid., p. 7.

5. Ibid., p. 8.

6. Ibid., pp. 9–10. Three references were not included in my count: one to the *Framework* document, one to the *National Science Education Standards*, and one to *Benchmarks for Science Literacy*.

7. "Topic Arrangements of the Next Generation Science Standards," p. 8. This document can be downloaded as a PDF file at: http://www.nextgenscience.org/sites/default/ files/NGSS%20Combined%20Topics%2011.8.13.pdf (accessed February 6, 2016).

8. Ibid., p. 12.

9. Ibid.

10. Ibid.

11. Ibid.

12. Ibid., p. 17.

13. Ibid., p. 38.

14. Ibid.

15. Ibid., p. 70.

16. Ibid., p. 68.

17. Ibid.

18. Ibid., p. 22.

19. Ibid., p. 27.

20. Ibid., p. 34.

21. Ibid., p. 22.

22. I did not count words that appeared in the white areas below the colored areas of the tables, which are essentially footnote areas.

23. "Topic Arrangement of NGSS," p. 85.

24. Ibid., pp. 68–69.

25. "Appendix K—Model Course Mapping in Middle and High School for the Next Generation Science Standards," p. 9, http://www.nextgenscience.org/sites/ngss/files/Appendix%20K_Revised%208.30.13.pdf (accessed January 23, 2016). There is a similar flow-diagram, fig. 4, on page 18.

26. Michigan State Board of Education, "Biology," *High School Content Expectations: Science*, p. 1, October 2006, http://detroit.k12.mi.us/admin/academic_affairs/science/docs/HSCE_Biology.pdf (accessed June 8, 2016). This particular document pertains to biology, but the first nine pages are common to each of the four sciences.

27. Ibid., p. 8.

28. Ibid.

29. Ibid.

30. Ibid., p. 2.

31. Ibid., p. 3.

32. Ibid., p. 4.

33. The learning pyramid appears on page after page of articles on the Internet, sometimes in the context of a useful way of illustrating a point in teaching and learning, and often as an object of derision. For an example of each see:

Saranne Magennis and Alison Farrell, "Teaching and Learning Activities: Expanding the Repertoire to Support Student Learning," in *Emerging Issues in the Practice of University Learning and Teaching*, ed. Geraldine O'Neill, Sarah Moore, and Barry McMullin (Dublin, Ireland: AISHE, 2005), http://www.aishe.org/readings/2005-1/magennis.html (accessed January 28, 2016).

Candice Benjes-Small, "Tales of the Undead . . . Learning Theories: The Learning Pyramid," *ACRLog*, January 13, 2014, http://acrlog.org/2014/01/13/tales-of-the-undead-learning-theories-the-learning-pyramid/ (accessed January 29, 2016).

34. Michigan State Board of Education, "Biology," p. 4.

35. Ibid., p. 5.

36. Oregon Department of Education, "Science Teaching and Learning to Standards: Oregon Teacher Resources 2007–2009," p. 9, http://www.ode.state.or.us/teachlearn/subjects/science/resources/sci-tls200709.pdf (accessed January 30, 2016).

37. Ibid., p. 11, figure 1.

38. Ibid.

39. Oregon Department of Education, "Science Teaching and Learning to Standards," p. 14.

40. Ibid., p. 24.

41. Ibid., p. 25.

42. Ibid.

43. Ibid., p. 36.

44. The quotation regarding Dr. Marylou Dantonio was taken from her former website, which is no longer available. On her LinkedIn profile for her reflexology and Reiki studio, she states that "she integrates energy modalities such as subtle energy work,

reflexology, chakra work, meridian reflexology, reiki, and intuiting with collegial reflective processes that she designed as a professional educator" (https://www.linkedin.com/in/marylou-dantonio-99221331 [accessed January 31, 2016]). Her joint publication with Paul Beisenherz is the book, *Using the Learning Cycle to Teach Physical Science: A Hands-On Approach to the Middle Grades*, published by Heinemann in 1996.

45. Oregon Department of Education, "Science Teaching and Learning to Standards," pp. 39–42. Design Space is credited to Dr. Dave Hamilton, Science Assessment Specialist for the Portland Public Schools, who has a doctorate in education.

46. Celeste Baine, *Teacher's Guide to Using Engineering Design in Science Teaching and Learning: Middle School Edition* (Portland, OR: Oregon University System, 2013), p. 4, http://www.ode.state.or.us/wma/teachlearn/edosc/teachers-guide-ms.pdf (accessed January 31, 2016).

47. Ibid., p. 10.

48. Ibid., p. 14.

49. Ibid.

50. Mary McClellan and Cary Sneider, *Washington State K–12 Science Learning Standards: Version 1.2 June 2010* (Olympia, WA: Office of Superintendent of Public Instruction, 2010), p. 2, http://www.k12.wa.us/science/pubdocs/WAScienceStandards.pdf (accessed February 3, 2016).

51. Ibid., p. 112. The full definition given is: "An abstract, universal idea of phenomena or relationships among phenomena."

52. Ibid., p. 123.

53. Ibid., p. 118.

54. Ibid., p. 110.

55. Ibid., p. 2.

56. Ibid., p. 4.

57. Ibid., p. 3.

58. Ibid., p. 4.

59. Ibid., pp. 106–109.

60. Ibid., p. 3.

61. Ibid.

62. Ibid., p. 2.

63. Ibid., p. 45.

64. Ibid.

65. Ibid., p. 63.

66. Ibid., p. 84.

67. Ibid., p. 49.

68. Ibid.

69. Ibid., p. 66.

70. The cord can be made by linking together a chain of rubber bands, or by using a length of braided elastic. By connecting the spring balance to the rubber cord by a short length of string tied in a loop, the cord could be released at will simply by cutting the string with scissors. The obvious links can be made to the storage of strain energy in a flexed bow prior to firing an arrow.

71. McClellan and Sneider, Washington State K–12 Science Learning Standards, p. 107.

72. Christopher McGowan, *A Practical Guide to Vertebrate Mechanics* (Cambridge: University of Cambridge Press, 1999), pp. 67 and 78.

73. McClellan and Sneider, *Washington State K–12 Science Learning Standards*, p. 115.

74. Ibid., p. 50.

75. Ibid., p. 44.

76. Ibid.

77. Ibid., p. 60.

78. "NGSS Writing Team Leader: Cary Sneider," Next Generation Science Standards, http://www.nextgenscience.org/ngss-writing-team-leader-cary-sneider (accessed October 17, 2016).

79. McClellan and Sneider, *Washington State K–12 Science Learning Standards*, p. 60.

80. "Topic Arrangement of NGSS," p. 55.

81. "Appendix K," p. 2.

82. "Appendix H," p. 2.

CHAPTER 8: THE NUFFIELD PROJECT

1. The Dutchman Antonie van Leeuwenhoek (1632–1723) was the first to report seeing bacteria (he referred to them as animalcules, from the Latin for *little animals*), using a simple microscope that he had built. During his lifetime he is said to have built hundreds of microscopes, some with magnifications of over 200 times, but few have survived. These microscopes used a single lens, which he made himself, and were not the forerunner of the compound microscopes used today.

2. The Pasteur experiment with bottles works well, but sometimes the milk in the plugged bottle also turns sour. The same thing was noticed during Pasteur's time when people attempting to repeat his original experiment occasionally failed to get the same results. The dilemma was explained in 1877 by John Tyndall, a prominent British physicist. Tyndall discovered that while bacteria are killed by boiling, the spores of some species are resistant to heat (spores are the reproductive stages that can remain dormant in harsh conditions until favorable conditions return). He devised a method of sterilization that overcame the problem. Known as *tyndallization*, the process involves heating things to boiling point on three successive days. Spores surviving the heating phase that go on and germinate into active bacteria during the cool period are subsequently destroyed during the next heating phase.

3. Most nematodes are parasitic, living inside the bodies of other animals, often causing diseases.

4. A human red blood cell (erythrocyte) has a mean diameter of 7.7 micrometers (thousandths of a millimeter).

5. To be correct, these are actually called antennules.

CHAPTER 9: THE WAY FORWARD

1. "Topic Arrangements of the Next Generation Science Standards," Next Generation Science Standards, November 2013, pp. 93 and 63, http://www.nextgenscience.org/sites/default/files/NGSS%20Combined%20Topics%2011.8.13.pdf (accessed February, 6, 2016).

2. Ian Abrahams and Robin Millar, "Does Practical Work Really Work? A Study of the Effectiveness of Practical Work as a Teaching and Learning Method in School Science," *International Journal of Science Education* 30, no. 14 (2008): 1945–69, http://citeseerx.ist.psu.edu/viewdoc/download?doi=10.1.1.456.354&rep=rep1&type=pdf (accessed April 3, 2016).

3. Ibid., p. 1946.

4. Ibid., p. 1954.

5. Ibid., p. 1955.

6. Ibid., p. 1956.

7. Ibid.

8. Ibid., p. 1957.

9. Two of the teachers were observed, on separate occasions, conducting two separate lessons; ibid., p. 1952, table 5.

10. Ibid., p. 1960.

11. Ibid., p. 1954.

12. Ibid., p. 1965.

13. J. T. Stock and P. Heath, *Small-Scale Inorganic Qualitative Analysis* (London: University Tutorial Press, 1954). I was surprised to see that this book is still available on the Amazon site.

14. "What are Learning Outcomes," Dean of Students, Brigham Young University, https://deanofstudents.byu.edu/content/what-are-learning-outcomes (accessed March 12, 2016).

15. This quote is taken from the *Handbook for Institutions Seeking Reaffirmation*, produced by the Southern Association of Colleges and Schools Commission on Colleges, August 2011 edition. The document is available at http://www.sacscoc.org/pdf/081705/Handbook%20for%20Institutions%20seeking%20reaffirmation.pdf (accessed March 13, 2016).

16. Christopher McGowan, *A Practical Guide to Vertebrate Mechanics* (Cambridge: Cambridge University Press, 1999).

17. Letter from Dr. Jennifer Keyte, DVM, Director of Animal Care Services, University Veterinarian, Memorial University, Newfoundland and Labrador, August 5, 2003.

18. To find out more about *Project Exploration* visit: http://www.projectexploration.org/about-us/ (accessed March 14, 2016).

19. White House, Office of the Press Secretary, "President Honors Outstanding Science, Math, Engineering Teachers and Mentors," news release, July 9, 2009, https://www.whitehouse.gov/the-press-office/president-honors-outstanding-science-math-engineering-teachers-and-mentors (accessed July 6, 2016).

20. Unfortunately, the hands-on Nuffield approach to teaching science has long since disappeared from UK classrooms, where educationalists have taken control, as they have in North America.

21. White House, Office of the Press Secretary, "Remarks by the President on the 'Education to Innovate' Campaign," transcript, November 23, 2009, https://www.whitehouse.gov/the-press-office/remarks-president-education-innovate-campaign (accessed July 6, 2016).

22. This particular American Chemical Society website is at: http://www.acs.org/content/acs/en/volunteer/chemambassadors/national-lab-day.html (accessed March 15, 2016).

23. The nonworking link of the website where I signed up for the National Lab Day project was at: https://www.nationallabday.org/projects/674-inquiry-experiences-in-science (first accessed in January 2010).

24. Michael S. Henry, "DOE Hosts National Lab Day on Capitol Hill," American Institute of Physics, July 10, 2015, https://www.aip.org/fyi/2015/doe-hosts-national-lab-day-capitol-hill (accessed March 15, 2016).

25. National Lab Network, http://www.nationallabnetwork.org/ (accessed March 15, 2016).

26. "National Lab Day," Facebook, https://www.facebook.com/NationalLabDay/ (accessed March 17, 2016).

CHAPTER 10: A CLASSROOM CHARTER

1. David Staples, "Suspended Teacher Lynden Dorval Is a Hero for Standing up for High Standards in the Classroom," *Edmonton Journal*, May 31, 2012, http://edmontonjournal.com/news/local-news/teacher-suspended-after-he-upheld-high-standards-is-a-hero (accessed July 6, 2016).

2. Charles F. Webber et al., *The Alberta Student Assessment Study Final Report* (Edmonton, Alta: Alberta Education, Learner Assessment, 2009), p. 12, https://archive.education.alberta.ca/media/1165612/albertaassessmentstudyfinalreport.pdf (accessed March 26, 2016).

3. Ibid., p. 48.

4. Michael Zwaagstra, "No-Zero Policies Just as Misguided as Ever," *Telegram*, January 14, 2015, http://www.thetelegram.com/Opinion/Letter-to-the-editor/2015-01-14/article-4006254/No-zero-policies-just-as-misguided-as-ever/1 (accessed March 26, 2016).

5. Webber et al., *Alberta Student Assessment Study*, p. 48.

6. Ibid.

7. Ibid., p. 135.

8. Ibid., p. 140.

9. Ibid.

10. Ibid.

11. Ibid.

12. The quote from Lynden Dorval is taken from Staples, "Suspended Teacher."

13. "Lynden Dorval, Fired for Giving Zeros, 'Treated Unfairly,' Appeal Board Rules," CBC News, August 29, 2014, http://www.cbc.ca/news/canada/edmonton/lynden-dorval-fired-for-giving-zeros-treated-unfairly-appeal-board-rules-1.2751007 (accessed July 6, 2016).

14. Gina Caneva, "For Students' Sake, Say No to 'No-Zero Policy' on Grading," *Catalyst Chicago*, October 22, 2013, http://catalyst-chicago.org/2013/10/students-sake-say-no-no-zero-policy-grading/ (accessed July 6, 2016).

15. OnTrack, http://ontrack.uchicago.edu/ (accessed July 6, 2016).

16. Caneva, "For Students' Sake," p. 3.

17. "Lowndes Co.' Grading Policy Makes National Headlines," WALB, February 6, 2012, http://www.walb.com/story/16689243/lowndes-co-grading-policy-makes-national-headlines (accessed March 28, 2016).

18. T. Rees Shapiro, "Fairfax Schools Consider New Grading Policy that Would Eliminate Zeros," *Washington Post*, March 18, 2015, https://www.washingtonpost.com/local/education/fairfax-schools-considers-new-grading-policy-that-would-eliminate-zeros/2015/03/18/9dd615c2-cd1e-11e4-a2a7-9517a3a70506_story.html (accessed March 30, 2016).

19. Ibid.

20. Nadia Goodman, "James Dyson on Using Failure to Drive Success," *Entrepreneur*, November 5, 2012, http://www.entrepreneur.com/article/224855 (accessed March 31, 2016).

21. Ibid.

22. Minutes, City of Fairfax School Board Regular Meeting No. 2, October 5, 2015, http://fairfax.granicus.com/DocumentViewer.php?file=fairfax_58a4dd0b-251e-4267-bd17-155552818582.pdf (accessed March 31, 2016).

23. Carly Rolph, "No-Zero Policy to Be Implemented in South Carolina Schools," *Daily Caller*, February 12, 2016, http://dailycaller.com/2016/02/12/no-zero-policy-to-be-implemented-in-south-carolina-schools/ (accessed April 1, 2016).

24. David Maylish, "Inflating Grades Lowers the Bar in Md. Education," *Baltimore Sun*, September 25, 2015, http://www.baltimoresun.com/news/opinion/oped/bs-ed-grade-inflation-20150927-story.html (accessed April 1, 2016).

25. Victor Skinner, "Alabama School Implements 'No Zeros' Grading Policy," *EAG News*, January 30, 2015, http://eagnews.org/alabama-school-implements-no-zeros-grading-policy/ (accessed April 1, 2016).

26. Ben Velderman, "Tucson Teachers Tell Principal Her No-Zero Grading Policy Is a No-Go," *EAG News*, August 2, 2012, http://eagnews.org/tucson-teachers-tell-principal-her-no-zero-grading-policy-is-a-no-go/ (accessed April 1, 2016).

27. Amy Sherman, "Jeb Bush Says Orange County Schools Made It Impossible for Students to Receive Below a 50," *PolitiFact Florida*, November 20, 2014, http://www.politifact.com/florida/statements/2014/nov/20/jeb-bush/jeb-bush-says-orange-county-schools-made-it-imposs/.

28. "Zero Tolerance," *Encyclopedia*, 2005, http://www.encyclopedia.com/topic/Zero_Tolerance.aspx (accessed April 2, 2016). (Original reference from Jeffrey Lehman and Shirelle Phelps, *West's Encyclopedia of American Law* (Detroit, Thomson/Gale, 2005.); Drug-Free Schools and Communities Act Amendments of 1989, Pub. L. No. 101–226, 103 Stat. 1928 (1989), https://www.gpo.gov/fdsys/pkg/STATUTE-103/pdf/STATUTE-103-Pg1928.pdf (accessed April 2, 2016).

29. "Sec. 4141. Gun-Free Requirements," US Department of Education, http://www2.ed.gov/policy/elsec/leg/esea02/pg54.html (accessed October 15, 2016).

30. Dennis Cauchon, "Zero-Tolerance Policies Lack Flexibility," *USA Today*, August 13, 1999, http://usatoday30.usatoday.com/educate/ednews3.htm (accessed April 2, 2016).

31. Eleanor J. Bader, "Paddles, Stun Guns and Chemical Sprays: How US Schools Discipline Students," *AlterNet*, March 16, 2016, http://www.alternet.org/education/paddles-stun-guns-and-chemical-sprays-how-us-schools-discipline-students (accessed April 3, 2016).

32. Carly Berwick, "Zeroing Out Zero Tolerance," *Atlantic*, March 17, 2015, http://www.theatlantic.com/education/archive/2015/03/zeroing-out-zero-tolerance/388003/ (accessed April 2, 2016).

33. National Center for Education Statistics, "Table 169: Number of Students Suspended and Expelled from Public Elementary and Secondary Schools, by Sex, Race/Ethnicity, and State: 2006," *Digest of Education Statistics*, 2006, http://nces.ed.gov/programs/digest/d10/tables/dt10_169.asp (accessed April 3, 2016).

34. Max Smith, "Fewer Students Expelled after Fairfax County Ends Zero-Tolerance Policy," WTOP, May 6, 2015, http://wtop.com/virginia/2015/05/fewer-students-expelled-fairfax-county-ends-zero-tolerance-policy/ (accessed April 3, 2016).

35. "New School Policy Changes How Bad Behavior Is Punished," *Your4State*, August 2, 2012, http://www.your4state.com/news/news/new-school-policy-changes-how-bad-behavior-is-punished (accessed April 3, 2016).

36. Paul Sperry, "How Liberal Discipline Policies Are Making Schools Less Safe," *New York Post*, March 14, 2015, http://nypost.com/2015/03/14/politicians-are-making-schools-less-safe-and-ruining-education-for-everyone/.

37. Ibid.

38. Laura Clark, "English Pupils 'Are Among the Worst Behaved in the World': Even Top-Rated Schools Are Blighted by Classroom Chaos, Study Says," *Daily Mail Online*, April 13, 2014, http://www.dailymail.co.uk/news/article-2603879/English-pupils-worst-behaved-world-Even-rated-schools-blighted-classroom-chaos-study-says.html (accessed April 4, 2016).

39. Terry Haydn, "To What Extent Is Behaviour a Problem in English Schools? Exploring the Scale and Prevalence of Deficits in Classroom Climate," *Review of Education* 2 (February 2014): 45–46, http://onlinelibrary.wiley.com/doi/10.1002/rev3.3025/epdf.

40. Robert Peal, "The Worst Behaved Pupils in the World? You'd Better Believe It: As a Study Says Schools Are Even More Anarchic than We Thought, the Shocking Testimony of a Once Idealistic Young Teacher," *Daily Mail Online*, April 24, 2014, http://www.dailymail.co.uk/news/article-2611750/The-worst-behaved-pupils-world-Youd

-better-believe-As-study-says-schools-anarchic-thought-shocking-testimony-idealistic-young-teacher.html (accessed April 4, 2016).

41. "What is Behaviour for Learning?" https://meagern.wordpress.com/what-is-behaviour-for-learning/.

42. Ibid.

43. Peal, "Worst Behaved Pupils in the World?"

44. Ibid.

45. "Summerhill: The Early Days," *A. S. Neill's Summerhill*, http://www.summerhillschool.co.uk/history.php (accessed April 7, 2016).

46. Ibid.

47. Ibid.

48. Ibid.

49. Matthew Appleton, *A Free Range Childhood: Self-Regulation at Summerhill School* (Burlington, VT: Resource Center for Redesigning Education, 2000); Ray Hemmings, *Fifty Years of Freedom: Study of the Development of the Ideas of A. S. Neill* (London: Allen & Unwin, 1972).

50. "Another Brick in the Wall, Part 2: Meaning," *Shmoop University, Inc.*, November 11, 2008, http://www.shmoop.com/another-brick-in-the-wall-part-2/meaning.html (accessed April 7, 2016).

51. "Another Brick in the Wall (Part II)," *Songfacts*, 2016, http://www.songfacts.com/detail.php?id=1696 (accessed April 7, 2016).

52. Pink Floyd, "Another Brick in the Wall Lyrics," *Pink-Floyd-Lyrics.com*, 2016, http://www.pink-floyd-lyrics.com/html/another-brick-in-the-wall-lyrics.html (accessed April 7, 2016).

53. Anna Dobbie, "Summerhill School Ten years after Victory Over OFSTED," BBC Radio Suffolk, June 16, 2010, http://news.bbc.co.uk/local/suffolk/hi/people_and_places/newsid_8743000/8743801.stm (accessed April 7, 2016).

54. A brief account of Summerhill's battles with the government inspectors, given on the school's excellent website, can be viewed at: "Summerhill's Fight with the UK Government," *A. S. Neill's Summerhill*, http://www.summerhillschool.co.uk/summerhills-fight.php (accessed April 8, 2016).

The government inspection reports can be accessed at the Ofsted (Office for Standards in Education) website, http://reports.ofsted.gov.uk/inspection-reports/find-inspection-report/provider/ELS/124870 (accessed April 8, 2016).

55. Peter Wilby, "Summerhill School: These Days Surprisingly Strict," *Guardian*, May 27, 2013, http://www.theguardian.com/education/2013/may/27/summerhill-school-head-profile.

56. Personal communication from Lynn, Summerhill School, April 11, 2016.

57. Wilby, "Summerhill School."

58. The Hackney riot was one of a dozen or more riots around the Greater London area, all sparked by the police shooting. They even extended to cities as far away as Liverpool and Bristol.

59. Jessica Shepherd and Jeevan Vasagar, "A-Level Results: University Places All

Around at School in Riot-Hit Area," *Guardian*, August 18, 2011, http://www.theguardian.com/education/2011/aug/18/a-levels-mossbournc-academy (accessed April 8, 2006).

60. Sharon Hendry, "We Tell Kids We Believe in Them and Give Them Love . . . But It's Tough Love," *Sun*, October 6, 2011, http://www.thesun.co.uk/sol/homepage/woman/parenting/3855905/We-tell-kids-we-believe-in-them-and-give-them-love-but-its-tough-love.html (accessed April 9, 2016).

61. Ibid.

62. Christopher Middleton, "The School That Beat the Rioters," *Telegraph*, August 16, 2011, http://www.telegraph.co.uk/education/secondaryeducation/8698216/The-school-that-beat-the-rioters.html (accessed April 9, 2016).

CHAPTER 11: AFTERMATH

1. Bob McDonald, in an email to the author on November 29, 2012.

2. Specifically for grades 9–11.

3. Letter to the Honourable Liz Sandals, Minister of Education, October 1, 2014.

4. Letter from the Honourable Liz Sandals, Minister of Education, November 18, 2014.

5. Ibid. The Honourable Liz Sandals has recently been replaced by a new Minister of Education, the Honourable Mitzie Hunter.

CHAPTER 12: THE REALLY IMPORTANT ISSUES

1. Spencer R. Weart, *The Discovery of Global Warming*, https://www.aip.org/history/climate/index.htm (accessed May 18, 2016).

2. J. G. J. Olivier et al., *Trends in Global CO_2 Emissions: 2015 Report* (The Hague: Netherlands Environmental Assessment Agency, 2015), http://edgar.jrc.ec.europa.eu/news_docs/jrc-2015-trends-in-global-co2-emissions-2015-report-98184.pdf (accessed April 23, 2016).

3. David E. Sanger, "Bush Will Continue to Oppose Kyoto Pact on Global Warming," *New York Times*, June 12, 2001, http://www.nytimes.com/2001/06/12/world/bush-will-continue-to-oppose-kyoto-pact-on-global-warming.html?pagewanted=all (accessed April 23, 2016).

4. Carol Linnitt, "Harper's Attack on Science: No Science, No Evidence, No Truth, No Democracy," *Academic Matters*, May 2013, http://www.academicmatters.ca/2013/05/harpers-attack-on-science-no-science-no-evidence-no-truth-no-democracy/.

5. Olivier et al., *Trends in Global CO_2 Emissions*.

6. White House, Office of the Press Secretary, "Remarks by the President at UN Climate Change Summit," transcript, September 23, 2014, http://www.whitehouse.gov/the-press-office/2014/09/23/remarks-president-un-climate-change-summit.

7. Ibid.

8. Coral Davenport, "Nations Approve Landmark Climate Accord in Paris," *New York Times*, December 12, 2015, http://www.nytimes.com/2015/12/13/world/europe/climate-change-accord-paris.html (accessed April 24, 2016).

IANS, "India Committed to Fight Global Warming: Javadekar Signs Paris Agreement," *First Post*, April 23, 2016, http://www.firstpost.com/politics/climate-change-paris-agreement-javadekar-india-global-warming-united-nations-2745296.html?utm_source=FP_CAT_LATEST_NEWS (accessed July 7, 2016).

9. Davenport, "Nations Approve Landmark Climate Accord."

10. Ibid.

11. National Research Council, *Surface Temperature Reconstructions for the Last 2,000 Years* (Washington, DC: National Academies Press, 2006), p. 30, http://oceanservice.noaa.gov/education/pd/climate/teachingclimate/surftemps2000yrs.pdf (accessed April 24, 2016). For a brief account of satellite imaging data see: "Sea Surface Temperature," Physical Oceanography Distributed Active Archive Center (PO.DAAC), Jet Propulsion Laboratory, California Institute of Technology, https://podaac.jpl.nasa.gov/SeaSurfaceTemperature (accessed April 25, 2016).

12. "Climate Change: How Do We Know?" *Global Climate Change: Vital Signs of the Planet*, NASA, http://climate.nasa.gov/evidence/.

13. Dana Bash and Deirdre Walsh, "Cruz to CNN: Global Warming Not Supported by Data," CNN, February 20, 2014, http://politicalticker.blogs.cnn.com/2014/02/20/cruz-to-cnn-global-warming-not-supported-by-data/ (accessed April 24, 2016).

14. John M. Broder, "Past Decade Warmest on Record, NASA Data Shows," *New York Times*, January 21, 2010, http://www.nytimes.com/2010/01/22/science/earth/22warming.html (accessed April 24, 2016).

15. Adam Voiland, "2009: Second Warmest Year on Record; End of Warmest Decade," Goddard Institute for Space Studies, NASA, January 21, 2010, http://www.giss.nasa.gov/research/news/20100121/ (accessed April 24, 2016).

16. Ibid.

17. Anthony Leiserowitz et al., *Climate Change in the American Mind, March 2015*, Yale Program on Climate Change Communication, April 20, 2015, http://climatecommunication.yale.edu/publications/global-warming-ccam-march-2015/; Lydia Saad, "US Views on Climate Change Stable after Extreme Winter," Gallup, March 25, 2015, http://www.gallup.com/poll/182150/views-climate-change-stable-extreme-winter.aspx; Gayathri Vaidyanathan, "Big Gap between What Scientists Say and Americans Think about Climate Change: But the Gap May be Closing between Scientists and the Public on Global Warming," *Scientific American*, January 30, 2015, http://www.scientificamerican.com/article/big-gap-between-what-scientists-say-and-americans-think-about-climate-change/ (all accessed April 25, 2016).

18. Art Swift, "Americans Split on Support for Fracking in Oil, Natural Gas," Gallup, March 23, 2015, http://www.gallup.com/poll/182075/americans-split-support-fracking-oil-natural-gas.aspx (accessed April 25, 2016).

19. A mud and water mix, looking something like chocolate milk, was used in the

early days of drilling, but a whole range of "drilling muds" are now available, ranging from synthetic oils to compressed air.

20. EPA, *Assessment of the Potential Impacts of Hydraulic Fracturing for Oil and Gas on Drinking Water Resources, Executive Summary* (Washington, DC: Office of Research and Development, June 2015), p. ES-5, https://www.epa.gov/sites/production/files/2015-07/documents/hf_es_erd_jun2015.pdf.

21. Ibid., p. ES-9.

22. Ibid., p. ES-6.

23. Ibid., p. ES-17.

24. The known or suspected carcinogens include benzene, 1,4-dioxane, 1-hexadecene, and 1-tetradecene. See, "Preliminary Revised Draft, Supplemental Generic Environmental Impact Statement on the Oil, Gas and Solution Mining Regulatory Program: Well Permit Issuance for Horizontal Drilling and High-Volume Hydraulic Fracturing to Develop the Marcellus Shale and Other Low-Permeability Gas Reservoirs," New York State Department of Environmental Conservation, September 30, 2009, http://energyindepth.org/wp-content/uploads/marcellus/2011/07/SGEIS-Preliminary-Revised-Draft-7-1-11.pdf (accessed October 18, 2016), pp. 5-54, 5-56 (pp. 215, 217 of the PDF file; this file contains 736 pages, numbered according to chapter; this information resides in chapter 5).

25. There are many accounts of the Halliburton Loophole. See, for example: Wenonah Hauter, "10 Years Later: Fracking and the Halliburton Loophole," *EcoWatch*, August 11, 2015, http://www.ecowatch.com/10-years-later-fracking-and-the-halliburton-loophole-1882083309.html (accessed July 7, 2016); "The Halliburton Loophole," editorial, *New York Times*, November 2, 2009, http://www.nytimes.com/2009/11/03/opinion/03tue3.html?_r=0 (accessed April 26, 2016); Renee Cho, "The Fracking Facts," *State of the Planet*, Earth Institute, Columbia University, June 6, 2014, http://blogs.ei.columbia.edu/2014/06/06/the-fracking-facts/ (accessed July 7, 2016).

26. Energy Policy Act of 2005: An Act to Ensure Jobs for Our Future with Secure, Affordable, and Reliable Energy, Pub. L. 109–58, 118 Stat. 594 (August 8, 2005), p. 102, http://energy.gov/sites/prod/files/2013/10/f3/epact_2005.pdf (accessed May 18, 2016).

27. A Senate proposal by a Democratic member to repeal the Halliburton loophole was defeated on January 28, 2015. With the single exception of Senator Marco Rubio of Florida, all Republican senators voted against the proposal, leaving the regulation of fracking firmly in the hands of state agencies and out of the jurisdiction of the EPA, just as the industry prefers. See: Mike Soraghan, "Hydraulic Fracturing: Senate Votes to Keep 'Halliburton Loophole'; Regulation Stays with States," *E&E Publishing*, January 29, 2015, http://www.eenews.net/stories/1060012514.

28. Abrahm Lustgarten, "Injection Wells: The Poison Beneath Us," *ProPublica*, June 21, 2012, https://www.propublica.org/article/injection-wells-the-poison-beneath-us (accessed April 30, 2016).

29. David Biello, "How Can We Cope with the Dirty Water from Fracking?" *Scientific American*, May 25, 2012, http://www.scientificamerican.com/article/how-can-we-cope-with-the-dirty-water-from-fracking-for-natural-gas-and-oil/ (accessed May 1, 2016).

Also see: Andrew Zaleski, "A Start-Up That's Solved Fracking's Dirty Problem,"

CNBC, February 17, 2015, http://www.cnbc.com/2015/02/17/a-start-ups-that-solved-frackings-dirty-problem.html (accessed May 1, 2016).

30. Nathaniel R. Warner, Cidney A. Christie, Robert B. Jackson, and Avner Vengosh, "Impacts of Shale Gas Wastewater Disposal on Water Quality in Western Pennsylvania," *Environmental Science & Technology* 47, no. 14 (2013): 11849–57, http://sites.nicholas.duke.edu/avnervengosh/files/2011/08/EST_impacts-of-shale-gas-wastewater.pdf.

31. Adam Sieminski, *Implications of the US Shale Revolution*, US Energy Information Administration, October 17, 2014, http://www.eia.gov/pressroom/presentations/sieminski_10172014.pdf, p. 2.

32. "How Much Carbon Dioxide Is Produced When Different Fuels Are Burned?" US Energy Information Administration, June 14, 2016, https://www.eia.gov/tools/faqs/faq.cfm?id=73&t=11.

33. "Overview of Greenhouse Gases," EPA, August 9, 2016, https://www3.epa.gov/climatechange/ghgemissions/gases/ch4.html.

34. "Number of Producing Gas Wells," US Energy Information Administration, July 29, 2016, http://www.eia.gov/dnav/ng/ng_prod_wells_s1_a.htm.

35. "A World-Leader in Wind Energy," denmark.dk: The Official Website of Denmark, November 2015, http://denmark.dk/en/green-living/wind-energy/ (accessed May 2, 2016).

36. Ibid.

37. Zachary Shahan, "Top Wind Power Countries per Capita (Clean Technica Exclusive)," *Clean Technica*, June 20, 2013, http://cleantechnica.com/2013/06/20/top-wind-power-countries-in-the-world-per-capita-per-gdp-in-total/ (accessed October 18, 2016).

38. "What Is US Electricity Generation by Energy Source?" US Energy Information Administration, April 1, 2016, https://www.eia.gov/tools/faqs/faq.cfm?id=427&t=3 (accessed May 2, 2016).

39. Allen McFarland, "Twelve States Produced 80% of US Wind Power in 2013," US Energy Information Administration, April 15, 2014, http://www.eia.gov/todayinenergy/detail.cfm?id=15851 (accessed May 2, 2016).

40. "Wind Energy's Frequently Asked Questions (FAQ)," EWEA: The European Wind Energy Association, http://www.ewea.org/wind-energy-basics/faq/ (accessed May 2, 2016).

41. Lindsay Wilson, "Average Household Electricity Use Around the World," *Shrink That Footprint*, http://shrinkthatfootprint.com/average-household-electricity-consumption (accessed May 2, 2016).

42. Jose Zayas et al., "Enabling Wind Power Nationwide," US Department of Energy, May 2015, http://www.energy.gov/sites/prod/files/2015/05/f22/Enabling-Wind-Power-Nationwide_18MAY2015_FINAL.pdf, pp. 9–10 (accessed May 2, 2016).

43. "Meet the New World's Biggest Wind Turbine," *Renewable Energy World*, February 4, 2014, http://www.renewableenergyworld.com/articles/2014/02/meet-the-new-worlds-biggest-wind-turbine.html (accessed May 2, 2016).

44. Ozzie Zehner, *Green Illusions: The Dirty Secrets of Clean Energy and the Future of Environmentalism* (Lincoln and London: University of Nebraska Press, 2012), pp. 51–58.

development/desa/population/publications/pdf/fertility/world-fertility-patterns-2015.pdf (accessed May 16, 2016).

74. Dambisa Moyo, *Dead Aid: Why Aid Is Not Working and How There Is a Better Way for Africa* (New York: Farrar, Straus and Giroux, 2009), p. 35.

75. Ibid., p. 48.

76. The TV interview with Bono can be seen on YouTube at: "Anderson Cooper Full Length Bono, K'naan Somalia Famine Interview," YouTube video, 12:36, posted by "deecee1000," August 10, 2011, https://www.youtube.com/watch?v=-i9tL8Q1Fxo (accessed May 17, 2016).

77. Shashank Bengali, "Bush Birth Control Policies Helped Fuel Africa's Baby Boom," McClatchy DC, December 13, 2009, http://www.mcclatchydc.com/news/nation-world/world/article24566695.html (accessed May 17, 2016).

78. "Country Comparison: Total Fertility Rate," *The World Factbook*, Central Intelligence Agency, https://www.cia.gov/library/publications/the-world-factbook/rankorder/2127rank.html (accessed May 17, 2016). The total fertility rates are given for 2015. While originally expressed to two decimal places, as they are in the UN World Population Prospects (the 2015 revision), when expressed to one decimal place, the rates are the same for both sources.

79. Amy Kazmin, "India's Efforts to Control Its Population Are Still Stuck in the Past," *Financial Times*, November 14, 2014, http://www.ft.com/intl/cms/s/0/1070d0b8-6bf9-11e4-b1e6-00144feabdc0.html#axzz48wl4iTl3 (accessed May 17, 2016). See also, "India: On the Path to Replacement-Level Fertility?" *World Population Data Sheet, 2011*, Population Reference Bureau, http://www.prb.org/publications/datasheets/2011/world-population-data-sheet/india.aspx (accessed May 17, 2016).

80. "Country Comparison: Total Fertility Rates."

81. "Analyses: Tables, Figures & Maps," China-Profile, June 9, 2011, http://www.china-profile.com/data/fig_WPP2010_TFR_1.htm (accessed May 17, 2016).

Also see "Tales of the Unexpected," *Economist*, July 11, 2015, http://www.economist.com/news/china/21657416-china-has-relaxed-its-one-child-policy-yet-parents-are-not-rushing-have-second-tales.

82. World Population Prospects: Key Findings and Advance Tables, 2015 Revision, Department of Economic and Social Affairs, Population Division, United Nations, Working Paper No. ESA/P/WP.241, p. 18, http://esa.un.org/unpd/wpp/publications/files/key_findings_wpp_2015.pdf (accessed May 16, 2016).

83. Thomas Robert Malthus, *An Essay on the Principle of Population* (London: J. Johnson, 1798), p. 4, available at http://www.esp.org/books/malthus/population/malthus.pdf.

84. Ibid., p. 7.

85. Ibid., p. 44.

86. Ibid.

87. Paul R. Ehrlich, *The Population Bomb* (New York: Ballantine Books, 1968).

88. Paul R. Ehrlich and Anne H. Ehrlich, "The Population Bomb Revisited," *The Electronic Journal of Sustainable Development* 1 (2009): 63.

89. Clyde Haberman "The Unrealized Horrors of Population Explosion," *New York Times* Retro Report, May 31, 2015, http://www.nytimes.com/2015/06/01/us/the-unrealized-horrors-of-population-explosion.html?_r=0.

90. Fred Pearce, "The Population Bomb: Has It Been Defused?" *Yale Environment 360*, August 11, 2008, http://e360.yale.edu/content/feature.msp?id=2042 (accessed May 16, 2016).

91. World Fertility Patterns 2015, p. 6.

92. World Population Prospects.

93. Haberman, "Unrealized Horrors."

94. "Topic Arrangements of the Next Generation Science Standards," Next Generation Science Standards, November 2013, p. 8, http://www.nextgenscience.org/sites/default/files/NGSS%20Combined%20Topics%2011.8.13.pdf (accessed February 6, 2016).

95. "Who Says Math Has to Be Boring?" *New York Times*, December 7, 2013, http://www.nytimes.com/2013/12/08/opinion/sunday/who-says-math-has-to-be-boring.html (accessed June 23, 2016).

96. The trend toward eliminating dissection in medical schools is happening in North America and the UK: Charlie Cooper and Lucy Anna Gray, "Lack of Anatomy Training Could Lead to Shortage of Surgeons," *Independent*, June 28, 2014, http://www.independent.co.uk/life-style/health-and-families/health-news/lack-of-anatomy-training-could-lead-to-shortage-of-surgeons-9570684.html (accessed July 10, 2016); Randy Dotinga, "Med Schools Cut Out Cadavers," *Wired*, May 19, 2003, http://www.wired.com/2003/05/med-schools-cut-out-cadavers/ (accessed July 10, 2016).

97. Jonida Tafilaku, "Science Students Need Lab Experience, but It's Nearly Impossible to Get," *Guardian*, July 8, 2014, https://www.theguardian.com/education/mortarboard/2014/jul/08/science-students-lab-work-experience-impossible (accessed July 10, 2016).

IMAGE CREDITS

Figure 1.1: The Provost and Fellows of Worcester College, Oxford
Figure 1.2: Drawn by Julian Mulock
Figure 1.3: Drawn by Julian Mulock
Figure 1.4: Photos by author
Figure 2.1: Drawn by Julian Mulock
Figure 2.2: Drawn by Julian Mulock
Figure 4.1: Public Domain
Figure 4.2: Courtesy of Special Collections and Archives, Wright State University
Figure 4.3: Use of image courtesy of David Ostrowski and John Eney
Figure 4.4: Use of image courtesy of David Ostrowski and John Eney
Figure 4.5: Drawn by Julian Mulock
Figure 4.6: Drawn by Julian Mulock
Figure 4.7: Drawn by Julian Mulock
Figure 4.8: Drawn by Julian Mulock
Figure 4.9: Public Domain
Figure 4.10: © Jim Almond, www.shropshirebirder.co.uk
Figure 4.11: Canadian Air Force
Figure 4.12: © Dennis W. Donohue / Shutterstock
Figure 4.13: Drawn by Julian Mulock
Figure 4.14: Drawn by Julian Mulock
Figure 4.15: Drawn by Julian Mulock
Figure 4.16: Drawn by Julian Mulock
Figure 4.17: © Jim Almond, www.shropshirebirder.co.uk
Figure 4.18: © Joel Blit / Shutterstock
Figure 4.19: © holbox/Shutterstock
Figure 4.20: Public Domain
Figure 4.21: Courtesy of Special Collections and Archives, Wright State University

Figure 4.22: Drawn by Julian Mulock
Figure 4.23: Wikimedia Creative Commons, Manfred Münch, Licensed under CC BY-SA 3.0
Figure 4.24: © Duade Paton, http://www.duadepaton.com
Figure 4.25: Drawn by Julian Mulock
Figure 4.26: Drawn by Julian Mulock
Figure 4.27: Drawn by Julian Mulock
Figure 4.28: Drawn by Julian Mulock
Figure 4.29: Photo by author
Figure 4.30: Royal Ontario Museum
Figure 4.31: Royal Ontario Museum
Figure 4.32: Royal Ontario Museum
Figure 4.33: Royal Ontario Museum
Figure 5.1: Photo by author
Figure 5.2: Drawn by Julian Mulock
Figure 5.3: Drawn by Julian Mulock
Figure 8.1: Photo by Félix Nadar Crisco
Figure 8.2: Drawn by Julian Mulock
Figure 8.3: Drawn by Julian Mulock
Figure 8.4: Photo by author
Figure 8.5: Drawn by Julian Mulock
Figure 8.6: Drawn by Julian Mulock
Figure 8.7: Drawn by Julian Mulock
Figure 8.8: Drawn by Julian Mulock
Figure 8.9: Drawn by Julian Mulock
Figure 8.10: Drawn by Julian Mulock
Figure 8.11: Photo by author
Figure 8.12: Photo by author
Figure 8.13: © Lebenkulturen.de/Shutterstock
Figure 12.1: Public Domain
Figure 12.2: Drawn by Julian Mulock
Figure 12.3: Drawn by Julian Mulock
Figure 12.4: Drawn by Julian Mulock
Figure 12.5: Drawn by Julian Mulock

INDEX